A Cosmic Journey

From the Big Bang to Black Holes

By Douglas Donovan

IBSN 979-8-8693-5945-2

First Edition 2025

Chapter 1: The Universe Revealed - An Introduction to Astrophysics

For thousands of years, humans have gazed upward at the night sky with wonder, curiosity, and an insatiable desire to understand our place in the cosmos. What began as simple observations of celestial patterns has evolved into one of humanity's most profound scientific endeavors: astrophysics. This field represents far more than casual stargazing or the romantic notion of contemplating distant worlds. Astrophysics is the rigorous application of physical laws and mathematical principles to understand the universe's most spectacular phenomena, from the nuclear furnaces burning within stars to the vast cosmic web that connects galaxies across billions of light-years.

The journey from ancient astronomical observations to modern astrophysics reveals humanity's evolving perspective on our cosmic significance. This transformation in understanding represents one of the most dramatic shifts in human thought, fundamentally altering how we perceive reality itself. Today, we stand at an extraordinary vantage point in cosmic history, equipped with tools and knowledge that allow us to peer back nearly to the universe's beginning and forward to contemplate its ultimate fate.

From Earth-Centered to Universe-Centered: A Cosmic Revolution

The story of astrophysics begins with a fundamental misconception that dominated human thinking for millennia. Ancient civilizations, despite their sophisticated mathematical and observational skills, naturally assumed that Earth occupied the center of all creation. This geocentric worldview seemed entirely reasonable based on daily experience—the Sun appears to revolve around Earth once per day, the Moon and planets follow their own paths while also seeming to orbit our planet, and the stars appear fixed on a celestial sphere rotating about Earth's axis.

The Greek mathematician Eudoxus, working around 400 BCE, developed one of the first systematic geocentric models using concentric spheres with Earth at the center. This framework was later refined by other prominent thinkers including Aristotle and Plato, whose philosophical arguments for cosmic perfection reinforced the geocentric perspective. The most sophisticated version emerged from Ptolemy's work, which served as the foundation for astronomical predictions and astrological charts for over 1,500 years.

Yet even in antiquity, alternative perspectives existed. The Greek scholar Aristarchus of Samos proposed a revolutionary heliocentric model around 280 BCE, correctly placing the Sun at the center with all known planets orbiting around it. However, this radical idea found few supporters and was largely forgotten, partly because it failed to address a crucial observational challenge: if Earth truly moved through space, astronomers should observe stellar parallax—the apparent shifting of star positions as Earth orbits the Sun.

The absence of detectable stellar parallax represented a significant obstacle to accepting heliocentrism. Ancient astronomers faced a choice: either the stars were immensely more distant than anyone imagined, making angular movements too small to detect, or Earth remained stationary. The simpler explanation seemed to support the geocentric model, and this reasoning prevailed for centuries.

The modern revolution began with Nicolaus Copernicus, whose 1543 work "De revolutionibus orbium coelestium" revived the heliocentric concept. Although Copernicus's calculations initially proved less accurate than existing geocentric models due to his assumption of circular orbits, his work planted seeds of doubt about Earth's centrality. The breakthrough came with Galileo Galilei's telescopic observations in 1609, which revealed Jupiter's moons—clear evidence that not everything orbited Earth.

This cosmic revolution culminated with our modern understanding that Earth orbits an ordinary star in the suburbs of a typical galaxy, which itself is one of hundreds of billions of galaxies scattered throughout an expanding universe. The progression from geocentric to heliocentric to galactocentric to cosmic perspectives represents one of humanity's greatest intellectual achievements, requiring us to abandon comfortable assumptions about our special place in creation while revealing the universe's true grandeur.

The Nature of Astrophysical Investigation

Modern astrophysics distinguishes itself from classical astronomy through its emphasis on understanding the physical processes that govern celestial phenomena. While traditional astronomy focused primarily on cataloging and predicting celestial motions, astrophysics seeks to explain why stars shine, how galaxies form and evolve, what powers the most energetic explosions in the universe, and how the cosmos itself came to exist.

This transformation from descriptive to explanatory science became possible through the recognition that the same physical laws operating on Earth also govern the behavior of distant cosmic objects. Isaac Newton's demonstration in the late 1600s that celestial bodies follow the same laws of physics as terrestrial objects marked a crucial philosophical advance. This principle, known as the Copernican principle, established that the universe operates according to consistent physical laws regardless of location.

The power of astrophysics lies in its ability to extract enormous amounts of information from the faintest signals reaching Earth from across the cosmos. Unlike other sciences that can conduct controlled laboratory experiments, astrophysicists must work as cosmic detectives, interpreting the messages that the universe sends through electromagnetic radiation, gravitational waves, and high-energy particles. Every photon of light carries information about the conditions where it originated, potentially encoding details about temperature, density, chemical composition, magnetic field strength, and motion of its source.

This detective work requires sophisticated understanding of how matter and energy interact under extreme conditions rarely encountered on Earth. Stellar cores reach temperatures of millions of degrees, neutron stars compress matter to nuclear density, and black holes create gravitational fields so intense that they warp spacetime itself. Astrophysics provides the theoretical framework and observational techniques necessary to decode these extreme environments and understand their role in cosmic evolution.

Light as the Cosmic Messenger

The foundation of astrophysical knowledge rests on our ability to analyze electromagnetic radiation from distant objects.

Light, in its broadest sense encompassing the entire electromagnetic spectrum, serves as the primary messenger carrying information across the cosmos. Understanding the nature and properties of electromagnetic radiation therefore becomes crucial for anyone seeking to comprehend how astrophysicists gather information about the universe.

Electromagnetic radiation consists of oscillating electric and magnetic fields propagating through space at the constant speed of light—approximately 300,000 kilometers per second in vacuum. This radiation spans an enormous range of wavelengths and frequencies, from radio waves with wavelengths of kilometers to gamma rays with wavelengths smaller than atomic nuclei. The electromagnetic spectrum provides a comprehensive toolkit for studying different aspects of cosmic phenomena, as each wavelength range reveals distinct physical processes.

Radio waves, at the long-wavelength end of the spectrum, can penetrate interstellar dust and reveal the structure of gas clouds, the magnetic fields of planets, and the emission from exotic objects like pulsars. Infrared radiation reveals the heat signatures of warm objects and can peer through dust to expose star-forming regions hidden at visible wavelengths. Visible light provides the familiar colors we associate with stars and galaxies, while ultraviolet radiation reveals the hottest and most energetic stellar atmospheres. X-rays and gamma rays require the most extreme conditions to produce, emerging from the vicinity of black holes, neutron stars, and the most powerful explosions in the universe.

The revolutionary aspect of modern astrophysics lies in combining observations across multiple wavelengths to construct comprehensive pictures of cosmic objects and phenomena. A single galaxy might appear dramatically different when observed in radio waves versus optical light versus X-rays, with each wavelength revealing different components and processes. Radio observations might show vast jets of particles streaming from the galaxy's central black hole, optical observations could reveal the stellar populations and spiral structure, while X-ray observations might expose hot gas and active star formation regions.

Spectroscopy represents perhaps the most powerful tool in the astrophysicist's arsenal, allowing detailed analysis of light to determine the physical conditions at its source. When light passes through a spectrometer, it separates into its component wavelengths, revealing a spectrum that serves as a "fingerprint" of the emitting or absorbing material. Each chemical element produces a unique pattern of spectral lines corresponding to specific electron transitions within atoms. By identifying these patterns, astronomers can determine not only which elements are present in distant stars and galaxies but also their abundances, temperatures, densities, magnetic field strengths, and velocities.

This spectroscopic analysis has revealed that the universe consists of the same chemical elements found on Earth, confirming the universality of physical laws while providing detailed insights into cosmic chemistry and evolution. The relative strengths and positions of spectral lines encode information about stellar atmospheres, interstellar clouds, galaxy dynamics, and even the expansion of the universe itself.

Measuring the Cosmic Scale

One of the most challenging aspects of astrophysics involves comprehending the enormous scales of space and time that characterize the universe. The distances involved stretch human intuition to its limits, requiring specialized units of measurement and conceptual frameworks to make sense of cosmic dimensions.

The fundamental unit for expressing astronomical distances is the light-year, defined as the distance light travels in one year. Given light's constant speed of approximately 300,000 kilometers per second, one light-year equals roughly 9.46 trillion kilometers. This unit proves invaluable not only for measuring distances but also for understanding the relationship between space and time in astronomical observations.

When we observe a star located 100 light-years away, we see it as it appeared 100 years ago because the light we detect left that star a century ago. This fundamental principle means that astronomy is inherently a historical science—the farther we look into space, the further back in time we peer. The most distant galaxies visible to our most powerful telescopes appear as they existed when the universe was young, providing direct observational windows into cosmic history.

The scale of cosmic distances defies easy comprehension. Our nearest stellar neighbor, Proxima Centauri, lies about 4.2 light-years away—a distance so vast that it would take our fastest spacecraft tens of thousands of years to reach. The Milky Way galaxy spans roughly 100,000 light-years, containing hundreds of billions of stars scattered across this enormous volume. The nearest major galaxy, Andromeda, lies about 2.5 million light-years away, and the observable universe extends roughly 46 billion light-years in all directions.

These cosmic scales reveal the universe's hierarchical structure, with each level containing objects vastly larger than the previous scale. Planets orbit stars, stars cluster into galaxies, galaxies group into clusters, and clusters connect through vast filaments and sheets in the cosmic web. Understanding this hierarchy proves crucial for appreciating how physical processes operating on small scales influence phenomena on cosmic scales.

The temporal scales involved in astrophysics are equally staggering. Stars like our Sun shine for billions of years, galaxies evolve over tens of billions of years, and the universe itself has existed for approximately 13.8 billion years. Many of the processes we observe in the universe unfold over timescales far exceeding human experience, requiring sophisticated theoretical models and computer simulations to understand their development and evolution.

The Universal Application of Physical Laws

Perhaps the most remarkable discovery in the history of science is that the same physical laws governing matter and energy on Earth also operate throughout the observable universe. This universality allows astrophysicists to apply terrestrial physics to understand phenomena occurring under conditions impossible to replicate in earthly laboratories.

The laws of thermodynamics explain how stars generate energy through nuclear fusion while gradually evolving through predictable life cycles. Newton's law of gravitation describes the orbital motions of planets, the structure of star clusters, and the large-scale architecture of the universe itself. Einstein's theories of relativity become essential for understanding black holes, neutron stars, and the expansion of spacetime. Quantum mechanics governs the behavior of matter under extreme conditions found in stellar cores and the early universe.

This universality extends beyond fundamental physics to encompass chemical processes, electromagnetic interactions, and even complex phenomena like shock waves and turbulence. The same hydrogen atoms found in Earth's oceans exist in distant galaxies, following identical quantum mechanical rules and producing the same characteristic spectral signatures. The nuclear reactions powering stars operate according to the same principles studied in terrestrial particle accelerators, though under conditions of temperature and pressure that far exceed anything achievable on Earth.

The consistency of physical laws across cosmic scales provides confidence in our understanding of distant and ancient phenomena. When we observe the light from a quasar billions of light-years away, we can interpret its spectrum using atomic physics developed through terrestrial experiments, knowing that the same quantum mechanical transitions that produce spectral lines in laboratory gases also operate in the extreme environments surrounding distant black holes.

The Cosmic Story Ahead

The universe revealed through modern astrophysics tells an epic story of cosmic evolution spanning nearly 14 billion years of history. This narrative begins with the Big Bang, when space, time, matter, and energy emerged from conditions beyond our current ability to describe. In the universe's first moments, fundamental particles and forces took their present forms, setting the stage for all subsequent cosmic development.

As the universe expanded and cooled, the first light elements formed through Big Bang nucleosynthesis, creating the hydrogen and helium that would fuel the first stars. These primordial stars, massive and short-lived, forged heavier elements in their cores and distributed them through explosive deaths, enriching the cosmos with the carbon, oxygen, and other elements necessary for complexity.

Gravitational attraction gradually assembled matter into the cosmic web of dark matter that provides scaffolding for galaxy formation. Within these galaxies, successive generations of stars continued the process of cosmic chemical evolution while planets formed around many stellar systems. On at least one planet orbiting an ordinary star in a typical galaxy, the complexity enabled by billions of years of stellar element production led to the emergence of life and eventually to conscious beings capable of contemplating their cosmic origins.

This story continues unfolding today as new stars form, galaxies collide and merge, and the universe itself evolves toward an uncertain future dominated by dark energy's mysterious influence. Understanding this cosmic narrative requires exploring the physics of stellar evolution, the nature of dark matter and dark energy, the formation of black holes and neutron stars, the processes that create and distribute heavy elements, and the ultimate fate awaiting our expanding universe.

The subsequent chapters of this book will explore each aspect of this cosmic story in detail, revealing how astrophysicists have assembled this remarkable narrative through careful observation, theoretical analysis, and the universal application of physical laws. From the quantum mechanics governing stellar cores to the general relativity describing black holes, from the nuclear physics of explosive stellar deaths to the cosmological principles governing universal expansion, we will trace the connections linking the smallest scales to the grandest cosmic phenomena.

The universe revealed through astrophysics proves far stranger and more magnificent than our ancestors could have imagined, yet it remains comprehensible through human reason and scientific investigation. This combination of cosmic grandeur and intellectual accessibility makes astrophysics one of humanity's most inspiring and rewarding scientific endeavors, offering profound insights into our origins, our place in the cosmos, and our ultimate destiny among the stars.

Chapter 2: Einstein's Revolution - Special Relativity and the Cosmic Speed Limit

At the dawn of the 20th century, physics appeared to be a mature and largely complete science. Newton's laws of motion and gravitation had successfully explained planetary orbits and terrestrial mechanics for over two centuries, while James Clerk Maxwell's equations beautifully described electromagnetic phenomena, including the nature of light itself. Yet beneath this apparent completeness lay troubling inconsistencies that would ultimately revolutionize our understanding of space, time, and the universe itself.

The resolution of these contradictions came through the brilliant insights of a young patent clerk named Albert Einstein, whose special theory of relativity fundamentally altered humanity's conception of physical reality. Far from being merely an abstract theoretical curiosity, Einstein's revelations about the nature of space and time prove essential for understanding the most energetic and exotic phenomena in the cosmos, from the nuclear furnaces powering stars to the extreme environments surrounding black holes.

The Crisis in Classical Physics

The stage for Einstein's revolution was set by a series of puzzling experimental results that classical physics could not adequately explain. The most significant of these emerged from attempts to measure Earth's motion through what 19th-century physicists believed to be the "luminiferous aether"—a hypothetical medium that was thought to permeate all space and serve as the carrier of light waves.

Just as sound waves require air to propagate and water waves need water, scientists reasoned that light waves must propagate through some medium. This aether was imagined to be an invisible, massless substance filling the universe, providing the reference frame in which light traveled at the speed predicted by Maxwell's equations. If Earth moved through this aether as it orbited the Sun, then the speed of light should appear different when measured in different directions relative to Earth's motion.

The definitive test of this idea came through the ingenious experiment designed by Albert Michelson and Edward Morley, conducted in the basement of a Western Reserve University dormitory during the summer of 1887. Their interferometer split a beam of light into two perpendicular paths, allowing incredibly precise measurements of any differences in light's travel time along different directions. If Earth were moving through the aether, light traveling parallel to this motion should take a different time to complete its journey than light traveling perpendicular to it.

The result was completely unexpected: no matter how carefully they measured or how they oriented their apparatus, Michelson and Morley found no evidence whatsoever for Earth's motion through any aether. The speed of light appeared to be exactly the same in all directions, regardless of Earth's orbital motion or rotation. This null result puzzled physicists for nearly two decades, leading to increasingly elaborate attempts to save the aether concept through theories involving aether drag or length contraction of the experimental apparatus.

Equally troubling was the theoretical prediction that emerged from Maxwell's equations themselves. These equations predicted a specific speed for electromagnetic waves in vacuum, but they provided no indication of the reference frame in which this speed should be measured. According to Newton's laws, velocities should add vectorially—if you throw a ball forward from a moving train, its speed relative to the ground equals the ball's speed relative to the train plus the train's speed relative to the ground. Yet Maxwell's equations seemed to suggest that light behaved differently, maintaining the same speed regardless of the motion of its source or observer.

This contradiction between electromagnetic theory and Newtonian mechanics created a fundamental crisis in physics. Either Maxwell's equations were incorrect, or Newton's seemingly universal laws of motion had limitations that no one had previously suspected. The resolution required a completely new understanding of the nature of space and time themselves.

Einstein's Revolutionary Postulates

Albert Einstein's breakthrough came not from new experimental data but from a radically different way of thinking about the fundamental assumptions underlying physics. In his landmark 1905 paper on special relativity, Einstein proposed that the puzzling results of electromagnetic experiments could be understood by accepting two simple but profound postulates.

Einstein's first postulate, known as the principle of relativity, states that the laws of physics are the same in all inertial reference frames. An inertial reference frame is simply one that moves at constant velocity without acceleration or rotation—essentially, any laboratory or coordinate system in which Newton's first law of inertia holds exactly. This postulate extends a principle that was already well-established for mechanical phenomena: you cannot tell whether you are moving at constant velocity just by performing experiments within your own reference frame.

Consider the familiar experience of riding in a smooth-moving train or airplane. When traveling at constant speed with no turbulence, you can perform any mechanical experiment—dropping a ball, pouring water, playing catch—and get exactly the same results as if you were stationary on the ground. Einstein's first postulate boldly extended this principle to all physical laws, including electromagnetic phenomena. There should be no experiment of any kind that can reveal absolute motion through space.

The second postulate proved far more revolutionary: the speed of light in vacuum is the same for all observers, regardless of their motion relative to the light source. This seemingly innocent statement contradicted centuries of intuition about how velocities combine. If someone on a train shines a flashlight forward, both the person on the train and a stationary observer on the platform will measure exactly the same speed for the light beam—not the speed of light plus the train's velocity, but simply the speed of light.

This postulate directly addressed the puzzling results of the Michelson-Morley experiment while eliminating the need for any aether medium. Light does not require a material substance to propagate; it travels through empty space at a universal constant speed that serves as a fundamental property of the universe itself. Maxwell's equations were correct, and the speed of light they predicted represents a universal cosmic speed limit that no material object or information can exceed.

These two postulates, simple as they appear, led to consequences that fundamentally altered our understanding of space and time. They revealed that what we perceive as separate entities—space and time—are actually interwoven aspects of a unified spacetime that can be stretched, compressed, and distorted by motion and gravity.

Time Dilation: When Clocks Disagree

One of the most startling implications of Einstein's postulates is that time itself is not absolute but depends on the relative motion between observers. This phenomenon, known as time dilation, means that moving clocks run slowly compared to stationary ones—an effect that becomes increasingly pronounced as speeds approach the speed of light.

The logic leading to time dilation follows directly from the constancy of light speed. Imagine a simple clock constructed from two mirrors separated by a fixed distance, with a pulse of light bouncing back and forth between them. Each round trip represents one tick of the clock. Now consider this clock as viewed by two different observers: one moving with the clock and another observing it from a stationary reference frame.

For the observer moving with the clock, the light pulse travels straight up and down between the mirrors, covering a distance of twice the mirror separation during each complete cycle. Since the speed of light is constant, this observer measures a specific time interval for each tick. However, for the stationary observer watching the moving clock, the situation appears quite different. Because the entire clock is moving horizontally, the light pulse must travel along a diagonal path to keep up with the mirrors' motion, covering a greater total distance during each cycle.

Since both observers must measure the same speed for the light pulse, yet the stationary observer sees it traveling a longer path, more time must elapse for each cycle as measured by the stationary observer. The moving clock therefore appears to run slowly compared to an identical stationary clock.

This effect is perfectly reciprocal due to the principle of relativity. From the perspective of the moving observer, the stationary clock appears to be the one in motion, and therefore appears to run slowly. Each observer sees the other's clock running slowly compared to their own—a situation that seems paradoxical but is perfectly consistent with the relativistic nature of simultaneity.

The mathematical relationship governing time dilation involves the Lorentz factor, typically represented by the Greek letter gamma (γ). For an object moving at velocity v relative to an observer, $\gamma = 1/\sqrt{(1 - v^2/c^2)}$, where c is the speed of light. At everyday speeds, this factor is essentially equal to one, making relativistic effects negligible. However, as speeds approach the speed of light, γ increases dramatically, leading to significant time dilation.

The famous "twin paradox" provides a concrete example of time dilation's implications. If one twin travels to a distant star at high speed while the other remains on Earth, the traveling twin will age more slowly and return younger than their Earth-bound sibling. This effect has been confirmed through numerous experiments, from observations of fast-moving subatomic particles to precision measurements using atomic clocks on high-speed aircraft and satellites.

Length Contraction: Space Gets Squeezed

Just as time dilates for moving objects, spatial dimensions also change with motion. Length contraction, also known as Lorentz contraction, describes how objects appear shortened along their direction of motion when viewed from a different reference frame. Like time dilation, this effect becomes significant only at speeds approaching the speed of light.

The mathematical relationship governing length contraction mirrors that of time dilation but operates in the opposite direction. While time intervals stretch for moving objects, spatial lengths contract by the same factor. If an object has a proper length L_0 when at rest, its length as measured by an observer moving relative to it becomes $L = L_0/\gamma$, where γ is the same Lorentz factor encountered in time dilation.

This contraction occurs only along the direction of motion; dimensions perpendicular to the velocity remain unchanged. A moving spacecraft would appear shorter from nose to tail but maintain its normal height and width. Like time dilation, length contraction is perfectly reciprocal—each observer sees the other's measuring rods contracted relative to their own.

The derivation of length contraction can be understood through its connection to time dilation. Consider measuring the length of a moving object by using synchronized clocks at both ends. If the observer measures the time it takes for light to travel from one end of the moving object to the other, the observed length can be calculated from the travel time. However, simultaneity itself becomes relative in Einstein's framework, complicating this seemingly straightforward measurement.

The mathematical relationship governing length contraction mirrors that of time dilation but operates in the opposite direction. While time intervals stretch for moving objects, spatial lengths contract by the same Lorentz factor. If an object has a proper length L_0 when at rest, its length as measured by an observer moving relative to it becomes $L = L_0/\gamma$, where γ is the same Lorentz factor encountered in time dilation. This means that $L = L_0\sqrt{(1 - v^2/c^2)}$, showing that the contracted length is always shorter than the proper length.

This contraction occurs only along the direction of motion; dimensions perpendicular to the velocity remain unchanged. A moving spacecraft would appear shorter from nose to tail but maintain its normal height and width. Like time dilation, length contraction is perfectly reciprocal—each observer sees the other's measuring rods contracted relative to their own. This reciprocity might seem paradoxical, but it reflects the fundamental relativity of simultaneity in Einstein's framework.

The effects of length contraction become dramatic only at extremely high velocities. At 10% the speed of light, the contraction is barely noticeable, reducing length by only about 0.5%. However, at 90% light speed, an object appears contracted to just 44% of its original length, and at 99.9% light speed, it shrinks to merely 4.5% of its proper length. These extreme contractions occur only in the most energetic astrophysical environments, where particles and plasma jets approach relativistic speeds.

One of the most elegant confirmations of both time dilation and length contraction comes from observations of cosmic ray muons reaching Earth's surface. Muons are unstable particles created when high-energy cosmic rays collide with atmospheric molecules about 10 kilometers above sea level. These particles have an extremely short lifetime of only 2 microseconds when measured at rest in laboratory conditions. Given their typical speeds of about 99.99% the speed of light, classical physics predicts they should travel only about 600 meters before decaying, far short of the 10-kilometer journey to Earth's surface.

Yet muons are routinely detected at sea level, providing compelling evidence for relativistic effects. From Earth's reference frame, time dilation allows the muons' internal clocks to run slow, extending their apparent lifetime by a factor of about 70. This gives them sufficient time to complete the journey to Earth's surface. From the muons' reference frame, the situation appears different but equally consistent with relativity. The muons experience their normal 2-microsecond lifetime, but the 10-kilometer thickness of Earth's atmosphere appears contracted to just 140 meters, allowing them to traverse this shortened distance before decaying.

Mass-Energy Equivalence: The Universe's Most Famous Equation

Perhaps no scientific equation is more widely recognized than Einstein's $E = mc^2$, yet few people fully appreciate its profound implications for understanding the universe. This deceptively simple relationship reveals that mass and energy represent different aspects of the same fundamental quantity, forever linking matter with the enormous energies that power stars, fuel cosmic explosions, and drive the most violent phenomena in the cosmos.

Einstein derived mass-energy equivalence as a direct consequence of special relativity's postulates, showing that any object possessing mass also possesses an equivalent amount of rest energy. The equation states that the rest energy E_0 of any object equals its rest mass m multiplied by the speed of light squared. The appearance of c^2 as the conversion factor between mass and energy reveals why even tiny amounts of matter contain enormous quantities of energy—the speed of light is approximately 300 million meters per second, making c^2 an extraordinarily large number.

This mass-energy relationship fundamentally differs from the classical physics assumption that mass and energy represent entirely separate quantities. In Einstein's framework, mass and energy are simply different manifestations of the same underlying physical reality, like two sides of the same coin. Whenever a system gains energy through any process, its mass increases correspondingly by $\Delta E/c^2$. Conversely, whenever a system loses energy, its mass decreases by the same factor.

The practical implications of mass-energy equivalence become apparent in nuclear processes, where small changes in mass produce enormous energy releases. During nuclear fusion in stellar cores, hydrogen nuclei combine to form helium, but the resulting helium nucleus has slightly less mass than the original hydrogen nuclei. This "mass defect" gets converted directly into energy according to $E = mc^2$, providing the power source that keeps stars shining for billions of years.

In nuclear fission, heavy nuclei like uranium-235 split into lighter fragments, again with a small but significant mass difference converted to energy. The energy released per nucleus in these reactions amounts to about 200 million electron volts, roughly 100 million times more energetic than typical chemical reactions involving the same number of atoms. This enormous energy release stems directly from the c^2 factor in Einstein's equation, demonstrating how mass-energy equivalence enables both stellar fusion and the destructive power of nuclear weapons.

Beyond nuclear processes, mass-energy equivalence governs phenomena throughout the universe. Every photon of light carries energy but has zero rest mass, existing as pure energy propagating at light speed. When matter and antimatter annihilate, they convert entirely into electromagnetic radiation, representing the complete transformation of rest mass into kinetic energy. Conversely, high-energy photons can create particle-antiparticle pairs, converting pure energy back into matter in accordance with Einstein's relationship.

The equation also explains why massive particles can never reach the speed of light. As an object's velocity approaches c, its relativistic energy $E = \gamma mc^2$ increases without bound as γ approaches infinity. Accelerating a massive object to light speed would require infinite energy, making c an absolute cosmic speed limit for anything possessing rest mass. Only massless particles like photons can travel at light speed, and they must always travel at exactly that speed.

Relativistic Phenomena in Astrophysics

Special relativity's effects, negligible in everyday terrestrial experience, become crucial for understanding the most energetic phenomena in the cosmos. From the precise redshifts of distant galaxies to the collimated beams emerging from black holes, relativistic physics provides the theoretical framework needed to decode the universe's most spectacular displays of matter and energy in motion.

The relativistic Doppler effect represents one of the most important applications of special relativity in astrophysics, allowing astronomers to measure the velocities of distant objects through careful analysis of their electromagnetic spectra. When a source of electromagnetic radiation moves relative to an observer, the observed frequency differs from the frequency emitted in the source's rest frame. For objects moving away from the observer, the radiation appears redshifted to lower frequencies, while objects approaching the observer produce blueshifted radiation at higher frequencies.

The relativistic Doppler formula differs significantly from its classical counterpart when dealing with objects moving at substantial fractions of light speed. For a source receding from the observer with velocity v, the observed frequency is $v' = v\sqrt{[(1 - v/c)/(1 + v/c)]}$, where v represents the frequency emitted in the source's rest frame. This relationship includes both the classical Doppler shift due to the changing distance between source and observer, and an additional time dilation factor that becomes increasingly important at high velocities.

Astronomers exploit the relativistic Doppler effect to study galaxy motions, stellar orbits around black holes, and the expansion of the universe itself. The systematic redshift of distant galaxies provided the first evidence for cosmic expansion, with more distant galaxies showing greater redshifts as they recede at higher velocities. Near supermassive black holes, stars can reach orbital velocities of several percent of light speed, producing measurable relativistic Doppler effects that confirm general relativity's predictions about strong gravitational fields.

Relativistic jets represent perhaps the most dramatic manifestation of special relativity in astrophysical settings. These collimated beams of plasma, launched from the vicinity of black holes and neutron stars, can extend for thousands of light-years while maintaining their coherent structure. The particles within these jets often travel at velocities exceeding 99% of light speed, creating spectacular displays of relativistic physics on cosmic scales.

The formation of relativistic jets involves complex interactions between magnetic fields, accretion disks, and rotating compact objects. Current models suggest that magnetic field lines threaded through hot accretion disk material can be twisted and amplified by the rotation of the central black hole or neutron star. These twisted magnetic fields then accelerate charged particles to enormous velocities, creating highly collimated outflows that maintain their focus across vast distances.

When relativistic jets point nearly toward Earth, they exhibit dramatic brightening due to relativistic beaming effects. Radiation emitted from material moving at high speed toward an observer becomes concentrated into a narrow cone, making the jet appear much brighter than it would if observed from the side. This beaming effect, combined with time compression due to the finite light travel time, can make relativistic jets appear to move faster than light speed—a purely observational illusion known as superluminal motion.

Active galactic nuclei, powered by supermassive black holes consuming surrounding matter, often display magnificent relativistic jets extending far beyond their host galaxies. These jets transport enormous amounts of energy into intergalactic space, heating the surrounding medium and influencing galaxy formation processes across cosmic time. The most powerful jets, found in objects called quasars and blazars, can outshine entire galaxies through their relativistic emissions.

On smaller scales, stellar-mass black holes in binary systems can produce relativistic jets when matter from a companion star spirals into their gravitational grasp. These "microquasars" serve as laboratories for studying relativistic jet physics in environments where changes occur on timescales of hours or days rather than millennia. Observations of microquasars have confirmed many theoretical predictions about jet formation while revealing unexpected complexities in these extreme physical processes.

Bridging to General Relativity

While special relativity successfully describes physics in the absence of gravitational fields, it cannot address one of the universe's most fundamental forces. Einstein recognized this limitation and spent the next decade developing general relativity, which extends relativistic physics to include gravity and curved spacetime. The transition from special to general relativity represents one of the most profound conceptual leaps in scientific history, transforming our understanding of space, time, and gravity itself.

The key insight leading toward general relativity emerged from Einstein's recognition that acceleration and gravity produce indistinguishable physical effects within sufficiently small regions of spacetime. This equivalence principle suggests that an observer in a windowless elevator cannot determine whether they feel a downward force due to gravity or upward acceleration of the elevator. The apparent equivalence between gravitational and inertial effects implied that gravity might not represent a force in the traditional sense, but rather a manifestation of spacetime geometry.

Einstein's equivalence principle extends beyond simple mechanical effects to encompass all physical phenomena. Light rays that travel in straight lines through empty space must bend when observed from an accelerating reference frame, just as objects follow curved trajectories under acceleration. If acceleration and gravity are truly equivalent, then light should also bend in gravitational fields—a prediction that general relativity would later confirm through precise observations of starlight bending around the Sun during solar eclipses.

The mathematical framework needed to describe curved spacetime requires sophisticated tools from differential geometry, extending far beyond the relatively simple algebra of special relativity. In curved spacetime, the familiar notion of straight lines gives way to geodesics—the closest approximation to straight lines possible in curved geometry. Massive objects follow geodesics through spacetime, with what we perceive as gravitational attraction actually representing the natural motion of objects along curved spacetime paths.

Einstein's field equations, the mathematical heart of general relativity, relate the curvature of spacetime to the matter and energy it contains. These equations reveal that massive objects warp the fabric of spacetime around them, with the degree of curvature depending on both the concentration of mass-energy and the distance from the source. This warping effect extends throughout the universe, creating a dynamic spacetime that responds to the distribution and motion of matter and energy.

The predictions of general relativity differ significantly from those of Newtonian gravity only in extreme conditions involving very massive objects, very high speeds, or very precise measurements. For most astronomical applications, Newton's simpler gravitational theory provides excellent approximations. However, near black holes, neutron stars, and other compact objects, general relativistic effects become crucial for understanding observed phenomena. The precise timing of pulsars, the orbital decay of binary systems through gravitational wave emission, and the extreme redshifts observed near black holes all require general relativity for proper interpretation.

Modern technology also depends critically on general relativistic corrections, particularly in the Global Positioning System (GPS). GPS satellites orbit Earth at altitudes where both special and general relativistic effects influence their onboard atomic clocks. Special relativistic time dilation causes the satellite clocks to run slow due to their orbital motion, while general relativistic effects cause them to run fast due to the weaker gravitational field at their altitude. The net effect requires GPS clocks to be adjusted by about 38 microseconds per day to maintain the system's accuracy.

As we prepare to explore the universe's most extreme environments in subsequent chapters, general relativity will provide the essential theoretical foundation for understanding black holes, neutron stars, and the large-scale structure of the cosmos itself. The marriage of special and general relativity creates a comprehensive framework for describing physical reality across all scales, from subatomic particles to the observable universe. Einstein's revolutionary insights about space, time, and gravity continue guiding our exploration of the cosmos, revealing new wonders with each generation of increasingly sophisticated observations and theoretical developments.

The journey from special relativity's seemingly simple postulates about the constancy of light speed and the equivalence of inertial reference frames leads inexorably toward profound questions about the nature of gravity, the structure of spacetime, and the ultimate fate of the universe. These questions will guide our exploration through the remaining chapters, as we discover how Einstein's insights illuminate everything from the nuclear processes powering stars to the cosmic expansion that shapes the universe's evolution on the grandest scales.

Chapter 3: The Expanding Universe - Discovery of Cosmic Evolution

The early decades of the twentieth century witnessed one of the most revolutionary discoveries in the history of science—the realization that our universe is not static but expanding. This discovery fundamentally transformed our understanding of cosmic history, revealing that the universe has evolved dramatically over billions of years and continues changing today. The story of this revelation illustrates how careful observation, theoretical insight, and technological advancement combine to reveal nature's deepest secrets, while also demonstrating how scientific breakthroughs can emerge from unexpected quarters.

The discovery of cosmic expansion represents far more than a simple astronomical observation. It opened the door to modern cosmology, provided the observational foundation for the Big Bang theory, and revealed that the universe itself has a history—complete with a beginning, an evolving middle, and an uncertain future. This chapter traces the remarkable journey from early hints of galactic motion through Edwin Hubble's groundbreaking measurements to today's precision cosmology, which has revealed the universe's accelerating expansion driven by mysterious dark energy.

The Cosmic Laboratory: Galaxies as Distant Beacons

Before exploring the universe's expansion, we must understand the cosmic objects that served as messengers carrying information about this grand phenomenon. In the early 1900s, astronomers observed numerous fuzzy patches of light scattered across the sky, which they called "spiral nebulae" due to their often spiral-shaped appearance. The nature of these objects remained hotly debated—were they nearby gas clouds within our own Milky Way galaxy, or were they "island universes" comparable in size to our entire galaxy but located at enormous distances?

The resolution of this "Great Debate" required advances in both observational technology and measurement techniques. The key breakthrough came through the work of Henrietta Swan Leavitt, who discovered that certain variable stars called Cepheids showed a precise relationship between their brightness and the period of their light variations. This period-luminosity relationship meant that Cepheid stars could serve as "standard candles"—objects whose intrinsic brightness was known, allowing astronomers to determine their distance by comparing apparent brightness with absolute brightness.

Edwin Hubble, working with the powerful 100-inch Hooker telescope at Mount Wilson Observatory, used Leavitt's period-luminosity relationship to measure distances to several spiral nebulae. His observations revealed that these objects lay far beyond the boundaries of the Milky Way, confirming that they were indeed separate galaxies comparable in size to our own. This discovery immediately expanded the known universe by factors of thousands, revealing that what astronomers had thought was the entire cosmos represented merely our local cosmic neighborhood.

The identification of galaxies as separate star systems provided the foundation for studying cosmic expansion. These galaxies, scattered throughout space at various distances, served as markers that could reveal the universe's large-scale behavior. Their enormous distances meant that light from these galaxies carried information about cosmic conditions from millions or billions of years ago, while their distribution throughout space provided a cosmic laboratory for testing theories about the universe's structure and evolution.

Hubble's Revolutionary Observations

With the extragalactic nature of spiral nebulae established, Edwin Hubble turned his attention to studying their motions through space. This investigation built upon earlier work by Vesto Slipher at Lowell Observatory, who had measured the spectra of numerous spiral nebulae and found that most showed redshifted spectral lines—a phenomenon initially interpreted as evidence that these objects were moving away from Earth.

Hubble combined Slipher's velocity measurements with his own distance determinations to create the first systematic study of the relationship between galactic distance and recession velocity. Using Cepheid variable stars to establish a distance scale and spectroscopic redshift measurements to determine velocities, Hubble analyzed a sample of 24 nearby galaxies and discovered a remarkable pattern: more distant galaxies appeared to be receding at higher velocities than nearby ones.

This relationship, published in Hubble's landmark 1929 paper, established what became known as Hubble's Law: the recession velocity of galaxies is proportional to their distance from us. Mathematically, this relationship can be expressed as $v = H_0 \times d$, where v represents the recession velocity, d is the distance to the galaxy, and H_0 is a constant of proportionality now called the Hubble constant. The value of this constant determines both the current expansion rate of the universe and, through theoretical models, the universe's age.

Hubble's initial measurements contained significant systematic errors—his distance scale was wrong by about a factor of seven, leading to age estimates for the universe that were uncomfortably young compared to geological estimates of Earth's age. Despite these calibration problems, the fundamental relationship between distance and velocity proved robust. Subsequent observations by Hubble and his colleague Milton Humason extended the survey to much greater distances, confirming and refining the original discovery while establishing it as a fundamental property of the cosmos.

The implications of Hubble's discovery were immediately recognized as revolutionary. If all galaxies appeared to be moving away from us, with more distant ones receding faster, this suggested that the universe itself was expanding uniformly in all directions. This expansion provided observational support for theoretical models developed by Alexander Friedmann, Georges Lemaître, and others, which had predicted cosmic expansion based on Einstein's general relativity.

Understanding Redshift: The Doppler Effect and Beyond

The interpretation of galactic redshift as evidence for cosmic expansion required careful consideration of the physical processes that could produce such systematic shifts in spectral lines. The most familiar mechanism for frequency shifts is the classical Doppler effect, where motion between source and observer causes apparent changes in wave frequency. However, the redshifts observed in distant galaxies, while initially interpreted through the Doppler effect, actually arise from a more fundamental phenomenon: the expansion of space itself.

Classical Doppler shift occurs when objects move through space relative to an observer. If a galaxy were simply moving away from us through static space, its light would indeed be redshifted according to the familiar Doppler formula. However, the systematic nature of cosmic redshift—affecting all distant galaxies equally regardless of their direction—suggested something more profound than simple motion through space.

The correct interpretation of cosmic redshift emerges from general relativity's description of expanding space. As the universe expands, the wavelengths of photons traveling through this expanding space get stretched along with the expansion. This "cosmological redshift" differs fundamentally from Doppler shift because it results from the changing geometry of spacetime rather than relative motion through fixed space. Galaxies aren't necessarily moving through space at the high velocities their redshifts might suggest; instead, space itself is expanding, carrying galaxies along with it.

This distinction becomes crucial for understanding extreme redshift values observed in the most distant galaxies. Some galaxies show redshifts corresponding to recession velocities exceeding the speed of light if interpreted as simple Doppler shifts. However, this apparent violation of special relativity resolves when we recognize that the galaxies aren't moving through space at these velocities—rather, space itself is expanding at rates that can exceed the speed of light without violating any physical principles.

The expansion of space provides a natural explanation for why the universe appears homogeneous and isotropic on large scales despite the apparent special position implied by having all galaxies recede from us. In an expanding universe, every observer sees the same pattern of recession velocities proportional to distance, regardless of their location. Just as every point on an expanding balloon's surface sees all other points receding with velocities proportional to their separation, every cosmic observer experiences the same Hubble expansion.

The Cosmic Distance Ladder: Measuring the Universe

Understanding cosmic expansion requires accurate distance measurements to galaxies across a wide range of distances. This challenge led to the development of the "cosmic distance ladder"—a sequence of overlapping measurement techniques that extends from our solar neighborhood to the observable universe's edge. Each rung of this ladder relies on methods that work reliably at specific distance ranges, with careful calibration ensuring consistency between different techniques.

The first rung of the cosmic distance ladder relies on direct geometric methods, primarily parallax measurements of nearby stars. As Earth orbits the Sun, nearby stars appear to shift position slightly relative to more distant background stars. This parallax effect provides direct distance measurements for stars within about 1,000 light-years of Earth, establishing absolute distances without relying on theoretical models or assumptions about stellar properties.

Parallax measurements provide the fundamental calibration for all subsequent distance determinations. By observing stars of known distance through parallax, astronomers can determine their absolute brightness and establish the intrinsic properties of various types of standard candles. This calibration proves essential for extending distance measurements far beyond the range of direct parallax observations.

The second rung involves main sequence fitting and variable star relationships, particularly the period-luminosity relationship of Cepheid variables discovered by Henrietta Leavitt. Cepheid stars pulsate with periods directly related to their absolute brightness—more luminous Cepheids have longer periods. By observing the period of a Cepheid variable and comparing its apparent brightness to the expected absolute brightness, astronomers can determine its distance with reasonable accuracy.

Cepheid variables can be detected and measured in nearby galaxies using large telescopes, extending distance measurements out to about 100 million light-years. The Hubble Space Telescope has been particularly valuable for Cepheid observations, as its high resolution allows detection of individual variable stars in galaxies far beyond the range of ground-based telescopes. These observations have been crucial for calibrating the Hubble constant and establishing the universe's expansion rate.

Beyond Cepheid range, astronomers rely on brighter standard candles including certain types of supernovae, surface brightness fluctuations in elliptical galaxies, and the luminosities of entire galaxies. Each method requires careful calibration against closer techniques, and systematic errors can accumulate as astronomers extend their reach to greater distances. Despite these challenges, the cosmic distance ladder has provided remarkably consistent results when different methods are properly calibrated.

Type Ia Supernovae: Cosmic Lighthouses

The discovery that fundamentally changed our understanding of cosmic expansion came through observations of Type Ia supernovae—stellar explosions that can briefly outshine entire galaxies while maintaining remarkably consistent peak luminosities. These "standard candle" properties make Type Ia supernovae ideal for measuring distances to galaxies billions of light-years away, far beyond the range of any other reliable distance indicator.

Type Ia supernovae occur in binary star systems where a white dwarf star gradually accumulates matter from a companion star. When the white dwarf's mass approaches the Chandrasekhar limit of about 1.4 solar masses, carbon fusion ignites throughout the star's interior, causing a thermonuclear explosion that completely disrupts the white dwarf. The remarkable consistency of this explosion mechanism—always occurring when white dwarfs reach approximately the same critical mass—explains why Type Ia supernovae show such uniform peak brightnesses.

The Phillips relationship, discovered in 1993, revealed that small variations in Type Ia supernova brightness could be corrected using the shape of their light curves. This correction allows astronomers to standardize these explosions as precision distance indicators, with accuracies approaching 7% for individual supernovae. The extreme brightness of these explosions—billions of times more luminous than the Sun—makes them visible across cosmological distances, providing unique probes of cosmic expansion throughout the universe's history.

In the 1990s, two independent research teams began systematic searches for distant Type Ia supernovae with the goal of measuring how cosmic expansion had changed over time. These projects—the Supernova Cosmology Project and the High-Z Supernova Search Team—expected to find that expansion was gradually slowing due to gravitational attraction between galaxies. The anticipated discovery was how much this deceleration had occurred, which would reveal the universe's total matter content and predict its ultimate fate.

The results proved completely unexpected. Instead of finding evidence for decelerating expansion, both teams discovered that distant supernovae appeared dimmer than predicted by models of decelerating expansion. This dimming indicated that these supernovae were farther away than expected, implying that cosmic expansion had actually accelerated over the past several billion years rather than slowed down. This discovery earned the 2011 Nobel Prize in Physics while fundamentally changing our understanding of the universe's composition and evolution.

The Accelerating Universe and Dark Energy

The discovery of cosmic acceleration revealed that our universe's expansion is not simply coasting from an initial Big Bang impulse but is actively being driven by some form of repulsive energy. This "dark energy" represents perhaps the greatest mystery in modern physics, comprising roughly 70% of the universe's total energy content while remaining completely unknown in terms of its fundamental nature.

The observational evidence for acceleration comes from comparing supernova distances with their redshifts across cosmic time. In a universe with constant expansion rate, this relationship would be perfectly linear. In a decelerating universe, distant (and therefore ancient) supernovae would appear closer than expected because expansion was faster in the past. However, observations show the opposite pattern—distant supernovae appear farther than expected, indicating that expansion was slower in the past and has since accelerated.

This acceleration appears to have begun roughly 5-6 billion years ago, when the universe's density had decreased sufficiently through expansion that dark energy's repulsive effects began dominating over matter's gravitational attraction. Prior to this transition, the universe's expansion was indeed decelerating as expected, but dark energy's increasing dominance has driven accelerating expansion ever since.

The nature of dark energy remains one of cosmology's most pressing questions. The simplest explanation involves Einstein's cosmological constant—a constant energy density inherent in space itself. This "vacuum energy" would provide steady repulsive pressure that becomes increasingly important as the universe expands and matter density decreases. However, theoretical calculations of vacuum energy disagree with observations by factors of 10^{120}, representing perhaps the worst prediction in physics history.

Alternative explanations for dark energy include quintessence—a dynamic energy field that can change over time—and modifications to general relativity that might mimic dark energy's effects. Recent observations from the Dark Energy Spectroscopic Instrument (DESI) suggest that dark energy's influence may indeed be evolving over time rather than remaining constant, potentially supporting quintessence models over the cosmological constant.

Modern Precision Cosmology

Today's understanding of cosmic expansion has evolved far beyond Hubble's initial observations, incorporating data from multiple independent methods and achieving precision that would have seemed impossible just decades ago. Modern measurements of the Hubble constant combine observations from the cosmic distance ladder, cosmic microwave background radiation, gravitational wave events, and other techniques to provide converging estimates of the universe's current expansion rate.

The cosmic microwave background, discovered in 1965 and mapped with exquisite precision by satellites like WMAP and Planck, provides an independent measure of cosmic expansion through its detailed temperature and polarization patterns. These patterns preserve information about the universe's geometry, expansion history, and composition from when it was only 380,000 years old, offering crucial constraints on cosmological parameters including the Hubble constant.

Large-scale galaxy surveys map the three-dimensional distribution of matter throughout the universe, revealing how cosmic expansion has affected structure formation over billions of years. Projects like DESI create detailed maps of galaxy positions and velocities across cosmic history, providing unprecedented precision in measuring how expansion has evolved over time. These surveys have confirmed the basic picture of accelerating expansion while revealing subtle effects that constrain theoretical models.

Gravitational wave detections provide yet another independent probe of cosmic expansion through observations of colliding neutron stars and black holes. When these events produce both gravitational waves and electromagnetic counterparts, astronomers can measure both the distance (from gravitational wave amplitude) and recession velocity (from electromagnetic redshift), providing "standard siren" measurements of cosmic expansion that don't rely on the traditional distance ladder.

Despite these advances, significant puzzles remain in our understanding of cosmic expansion. Different measurement methods yield slightly different values for the Hubble constant, creating the "Hubble tension" that may indicate missing physics in our cosmic models. Similarly, the fundamental nature of dark energy remains unknown, with recent hints that it may be evolving over time adding new complexity to cosmic evolution models.

Implications for Cosmic Time and Universal History

The discovery of cosmic expansion revolutionized our understanding of time and history on cosmic scales. An expanding universe implies a finite age—by extrapolating expansion backward, we can estimate when all matter was concentrated at effectively infinite density, marking the universe's beginning in the Big Bang. Current measurements indicate the universe is approximately 13.8 billion years old, providing a cosmic timeline against which all astronomical phenomena must be understood.

The concept of cosmic time emerges naturally from universal expansion, providing a universal clock that synchronizes across the cosmos. Events that occurred at the same cosmic time—measured as the same interval after the Big Bang—appear at the same redshift regardless of their location in space. This cosmic time frame allows astronomers to study the universe's evolution systematically, comparing objects at different redshifts to trace how cosmic structures have changed over billions of years.

Cosmic expansion also explains why the night sky is dark despite containing billions of stars and galaxies. In a static, infinite universe filled with stars, every line of sight should eventually encounter a star's surface, making the entire sky as bright as the Sun. However, expansion limits the observable universe to a finite region, while cosmological redshift shifts starlight to longer wavelengths, naturally explaining the darkness of the night sky through what's known as Olbers' paradox resolution.

The expansion's future evolution depends critically on dark energy's properties and how they may change over time. If dark energy remains constant, expansion will continue accelerating forever, eventually separating galaxies so far that future astronomers will see only their local galaxy group against an empty, dark sky. However, if dark energy weakens or changes sign, the universe might eventually recollapse in a "Big Crunch," or undergo other exotic evolutionary paths that remain largely unexplored.

Understanding cosmic expansion has also revealed the universe's remarkable homogeneity and isotropy on the largest scales, despite the obvious clumping seen in galaxies and galaxy clusters. This large-scale uniformity, combined with the precise thermal equilibrium observed in the cosmic microwave background, presents puzzles about how such uniform conditions could have been established in the early universe—questions that will be explored in the following chapter on Big Bang cosmology.

The expanding universe discovered through Hubble's observations and refined through decades of subsequent research represents one of science's greatest triumphs, revealing a dynamic cosmos with a finite age, an evolving structure, and an uncertain but fascinating future. This discovery provides the observational foundation for modern cosmology while raising profound questions about dark energy, cosmic inflation, and the ultimate fate of everything we can observe. As we prepare to explore the Big Bang's details in the next chapter, the expanding universe provides the crucial framework for understanding how our cosmos evolved from its earliest moments to the complex structure we observe today.

Chapter 4: The Big Bang - Origins of Space, Time, and Matter

The Big Bang theory stands as one of the most successful and profound scientific theories ever developed, describing nothing less than the origin and early evolution of everything we can observe in the universe. This remarkable framework explains how space, time, matter, and energy emerged from conditions so extreme that our current understanding of physics breaks down, then evolved through a series of well-defined epochs to produce the cosmos we inhabit today. The theory's success lies not merely in its elegant mathematical structure, but in its ability to make precise, testable predictions that have been confirmed through diverse and independent observations spanning decades of research.

Far from being mere speculation about unknowable ancient events, Big Bang cosmology represents a triumph of the scientific method, combining theoretical insights with observational evidence to reconstruct the universe's history with remarkable precision. The theory successfully predicts the abundances of the lightest chemical elements, the existence and properties of the cosmic microwave background radiation, the large-scale structure of matter in the universe, and numerous other phenomena that would remain mysterious without this unifying framework. Understanding the Big Bang is essential for comprehending not only how our universe began, but why it has the specific properties we observe today.

Conceptual Foundations: What the Big Bang Really Represents

Perhaps no scientific theory suffers from more widespread misconceptions than Big Bang cosmology, largely due to the unfortunate choice of terminology that has persisted since Fred

Hoyle coined the phrase in 1949 as a dismissive criticism. The term "Big Bang" conjures images of a massive explosion occurring at a specific location in empty space, hurling matter outward like cosmic shrapnel. This mental picture, while vivid and memorable, represents almost the opposite of what actually occurred according to modern cosmological understanding.

The Big Bang was not an explosion in space, but rather an explosion of space itself. There was no central point where the explosion occurred, no pre-existing void into which matter expanded, and no external reference frame from which the event could be observed. Instead, the Big Bang represents the simultaneous appearance and initial expansion of space, time, matter, and energy throughout the universe. Every point in space was equally the "center" of this expansion, just as every point on an expanding balloon's surface moves away from every other point without any single point being special.

This fundamental distinction between expansion in space versus expansion of space resolves many apparent paradoxes and misconceptions about Big Bang cosmology. The universe does not expand into anything because there is nothing "outside" the universe for it to expand into—the very concepts of space and distance are defined within the universe itself. Similarly, there is no edge or boundary to the expanding universe because space itself is expanding everywhere, not just at some distant frontier.

The physics governing cosmic expansion emerges directly from Einstein's general theory of relativity, which describes how the presence of matter and energy curves spacetime while curved spacetime tells matter how to move. When applied to the universe as a whole, Einstein's field equations predict that a uniform distribution of matter and energy cannot remain static—

it must either expand or contract under its own gravitational influence. The observed expansion of the universe represents the natural consequence of these fundamental physical laws operating on cosmic scales.

Understanding the Big Bang as spacetime expansion rather than a conventional explosion also clarifies the relationship between cosmic expansion and the apparent recession of galaxies observed by Hubble. Galaxies are not flying apart due to some initial explosive impulse; instead, they maintain essentially the same positions within expanding space while the distances between them increase as space itself grows larger. This expansion carries galaxies along passively, like raisins in expanding bread dough, rather than propelling them actively through static space.

Tracing Cosmic History Backward

The Big Bang theory emerges naturally from projecting the currently observed cosmic expansion backward through time, following the logical implications of general relativity combined with observational evidence for the universe's large-scale homogeneity and isotropy. If galaxies are currently separating due to cosmic expansion, then in the past they must have been closer together. Extrapolating this reasoning leads inevitably to a time when all observable matter was concentrated at enormous density and temperature—the initial singularity that marks the beginning of space and time as we know them.

This backward extrapolation reveals a universe that becomes progressively hotter, denser, and more energetic as we approach earlier epochs. The relationship between cosmic expansion and temperature arises because expanding space stretches electromagnetic radiation, cooling it in the process, while the reverse compression heats radiation to higher temperatures. As we trace cosmic history backward, we encounter successive thresholds where increasingly exotic physics becomes important for understanding the universe's behavior and evolution.

The earliest epoch accessible to current physical understanding begins roughly 10^{-43} seconds after the nominal "beginning," when quantum gravitational effects dominated cosmic evolution. Prior to this Planck time, the universe existed in states of such extreme density and energy that our current theories of physics—both general relativity and quantum mechanics—break down completely. The Planck epoch remains one of physics' great frontiers, requiring a successful theory of quantum gravity to understand.

Following the Planck epoch, the universe underwent a period of exponential expansion called cosmic inflation, during which space expanded at rates far exceeding the speed of light while temperatures remained extraordinarily high. This inflationary period, lasting perhaps 10^{-32} seconds, explains several otherwise puzzling features of the observable universe, including its large-scale homogeneity, geometric flatness, and the absence of exotic relics like magnetic monopoles predicted by grand unified theories.

The end of inflation marked the beginning of the hot Big Bang phase properly speaking, when inflationary energy converted into the matter and radiation that would eventually evolve into stars, galaxies, and everything else we observe today. This reheating process filled the universe with a hot, dense plasma of elementary particles and antiparticles in thermal equilibrium, setting the stage for the sequence of particle physics processes that shaped the universe's subsequent evolution.

The First Three Minutes: Nucleosynthesis and Light Element Formation

One of Big Bang cosmology's greatest triumphs involves its successful prediction of light element abundances throughout the universe, providing compelling evidence that the cosmos indeed experienced the extreme conditions described by the theory. Big Bang nucleosynthesis (BBN) occurred during the universe's first twenty minutes, when temperatures and densities were sufficient for nuclear fusion reactions to build light elements from individual protons and neutrons.

The nucleosynthesis process began when the universe had cooled sufficiently for protons and neutrons to overcome their mutual repulsion and bind together through the strong nuclear force. Initially, roughly equal numbers of protons and neutrons existed in thermal equilibrium, but as the universe cooled below about one billion degrees, neutron decay began reducing the neutron fraction. By the time nucleosynthesis began in earnest, the neutron-to-proton ratio had decreased to approximately one to seven.

The first nuclear reactions produced deuterium—heavy hydrogen consisting of one proton and one neutron bound together. However, deuterium nuclei remained vulnerable to destruction by high-energy photons until the universe cooled below the deuterium binding energy, creating a "deuterium bottleneck" that delayed significant nucleosynthesis until temperatures dropped to about 100 million degrees. Once this threshold was crossed, deuterium production proceeded rapidly.

Most deuterium nuclei quickly fused with additional protons or neutrons to form helium-3 and tritium, which then combined to produce helium-4—the second-lightest and second-most abundant element in the universe. The nuclear reaction chains were remarkably efficient at converting available neutrons into helium-4, with nearly every neutron present at nucleosynthesis onset ultimately incorporated into helium nuclei. This efficiency explains why helium-4 comprises about 25% of the universe's ordinary matter by mass.

Beyond helium, Big Bang nucleosynthesis produced only trace amounts of lithium and beryllium. The rapid expansion and cooling of the universe prevented the formation of heavier elements, which require either more complex reaction chains or higher densities than were available during the brief nucleosynthesis epoch. Elements heavier than lithium would have to await the formation of stars, where longer timescales and higher densities enable the nuclear processes that forge carbon, oxygen, and all heavier elements.

The predicted abundances of light elements depend sensitively on the density of ordinary matter (baryons) during nucleosynthesis. Higher baryon densities lead to more efficient nuclear reactions and different final element abundances, while lower densities reduce reaction rates and leave more free protons unconsumed. This density dependence allows astronomers to use observed light element abundances to determine the cosmic baryon density, providing an independent measurement that agrees remarkably well with values determined from cosmic microwave background observations.

The Cosmic Dark Age and Recombination

Following nucleosynthesis, the universe entered what cosmologists call the Dark Age—a period lasting several hundred thousand years when no stars or other luminous objects existed to illuminate the cosmos. During this epoch, the universe consisted primarily of hydrogen and helium gas continued from nucleosynthesis, entering a period that cosmologists call the Dark Age—a time when the universe contained no luminous objects and remained largely invisible to any hypothetical observer. During this epoch, the cosmos consisted primarily of hydrogen and helium gas mixed with energetic photons, all maintained at extremely high temperatures by the leftover thermal energy from the Big Bang itself.

The physics governing this period was dominated by the ongoing battle between the universe's expansion, which continuously cooled the cosmic material, and the intense radiation field that kept atoms from forming stable structures. Although nucleosynthesis had produced the raw materials for the first atoms, the ambient temperature remained far too high for electrons to bind permanently to atomic nuclei. Any electron that temporarily attached to a proton would quickly absorb a high-energy photon and escape, preventing the formation of stable hydrogen atoms.

This situation created a cosmic catch-22 that persisted for hundreds of thousands of years. The universe contained all the ingredients needed for transparency—protons, electrons, and photons—yet remained completely opaque because these components could not achieve stable configurations. The high-energy radiation environment meant that photons could travel only extremely short distances before encountering a free electron and scattering in a new direction, creating a cosmic fog that would have made the universe appear blindingly bright but completely featureless to any observer.

The Great Transition: Recombination and Cosmic Transparency

The cosmic Dark Age ended dramatically around 380,000 years after the Big Bang, when the universe finally cooled sufficiently for a process called recombination to occur. This somewhat misleadingly named event—electrons and protons had never been "combined" before, so "combination" would be more accurate—marked one of the most significant transitions in cosmic history, fundamentally altering the universe's physical properties and observational characteristics.

Recombination began when the cosmic temperature dropped below approximately 3,000 Kelvin, cool enough that the binding energy of hydrogen atoms could overcome the disruptive effects of ambient thermal radiation. At this critical temperature, electrons could finally attach to protons and remain bound long enough to form stable neutral hydrogen atoms, dramatically reducing the number of free electrons available to scatter photons. This process proceeded relatively quickly once it began, with the ionization fraction dropping from 90% to just 10% over roughly 70,000 years.

The physics of recombination proves more subtle than simple thermal equilibrium calculations might suggest. Direct recombination to hydrogen's ground state posed significant challenges because the resulting 13.6 electron volt photons could immediately ionize other hydrogen atoms, preventing net progress toward neutrality. Instead, recombination proceeded primarily through a two-step process where electrons first captured into excited hydrogen states before cascading down to the ground level, with the lower-energy photons from these transitions being less likely to cause reionization.

Helium recombination actually preceded hydrogen recombination, occurring in two stages due to helium's higher ionization energies. Doubly ionized helium (He^{2+}) first captured electrons to form singly ionized helium (He^{+}) around redshift $z \approx 6000$, followed by a second recombination stage around $z \approx 2000$ that produced neutral helium atoms. However, helium recombination had relatively little effect on cosmic opacity since hydrogen's much greater abundance dominated the free electron population.

The completion of hydrogen recombination transformed the universe from an opaque plasma into a transparent medium almost literally overnight on cosmic timescales. With the vast majority of electrons now bound in neutral atoms, photons could suddenly travel enormous distances without scattering, allowing light to stream freely through space for the first time in cosmic history. This transition created what cosmologists call the "surface of last scattering"—the epoch when photons last interacted significantly with matter before beginning their journey across the universe.

Birth of the Cosmic Microwave Background

The photons released during recombination constitute perhaps the most remarkable fossil in the universe: the cosmic microwave background (CMB) radiation that permeates all of space and provides a direct observational window into conditions when the cosmos was less than 400,000 years old. These ancient photons carry encoded information about the early universe's temperature, density, composition, and even the seeds of future cosmic structure, making the CMB one of cosmology's most powerful observational tools.

When first emitted during recombination, these photons possessed energies corresponding to visible and near-infrared light, with the universe glowing at roughly the temperature of a red giant star's surface. However, the subsequent expansion of space has stretched these photons' wavelengths by a factor of about 1,100, shifting them from optical frequencies into the microwave portion of the electromagnetic spectrum. Today, this radiation maintains a nearly perfect blackbody spectrum at a temperature of 2.7255 Kelvin, making it detectable with sensitive radio telescopes but invisible to human eyes.

The accidental discovery of the CMB in 1965 by Arno Penzias and Robert Wilson at Bell Telephone Laboratories represents one of the most important serendipitous findings in scientific history. Working with a sensitive horn antenna originally designed for satellite communications, they detected a persistent background noise that appeared to come from all directions in the sky with equal intensity. Initially interpreting this signal as equipment malfunction or interference, they spent months trying to eliminate the source, even cleaning pigeon droppings from their antenna, before recognizing they had discovered the cosmic microwave background predicted by Big Bang theory.

The CMB's properties provide compelling evidence for the hot Big Bang model while ruling out alternative cosmological theories that were once competitive. The Steady State theory, which proposed that the universe had always existed in roughly its current form with new matter continuously created to maintain constant density, could not explain the existence of this thermal background radiation. Only a universe that had evolved from a hot, dense initial state could account for the observed characteristics of the microwave background.

Modern observations have revealed that the CMB is not perfectly uniform but contains tiny temperature variations at the level of about one part in 100,000. These minute fluctuations represent the imprints of quantum mechanical variations in the early universe that were amplified by cosmic inflation and later evolved into the galaxies and large-scale structures we observe today. The pattern of these temperature variations across the sky encodes precise information about fundamental cosmological parameters, including the universe's age, geometry, expansion rate, and the relative abundances of ordinary matter, dark matter, and dark energy.

The CMB's detailed spectrum and angular distribution have been mapped with extraordinary precision by satellite missions including COBE, WMAP, and Planck. These observations have confirmed Big Bang predictions with remarkable accuracy while providing the foundation for precision cosmology that can determine fundamental parameters of the universe to better than 1% accuracy. The acoustic oscillation patterns visible in CMB temperature fluctuations reveal how pressure waves propagated through the early universe's photon-baryon plasma, providing direct evidence for the physical processes described by Big Bang theory.

Multiple Pillars of Evidence: The Big Bang's Observational Foundation

The strength of Big Bang cosmology lies not in any single observation but in the convergence of multiple independent lines of evidence that together create an overwhelmingly compelling case for this model of cosmic origins. Each piece of evidence could potentially be explained by alternative theories, but the combination of observations from completely different physical phenomena all pointing toward the same conclusion provides the kind of robust confirmation that characterizes mature scientific theories.

The expansion of the universe, first discovered through Hubble's observations of galactic redshifts, provides the foundational evidence that motivated Big Bang theory in the first place. This expansion is not merely consistent with an evolving universe but actually demands it according to the equations of general relativity applied to a homogeneous, isotropic cosmos. The precise relationship between galaxy distance and recession velocity, refined through decades of observations, matches the predictions of Big Bang models while constraining fundamental parameters like the Hubble constant and the universe's age.

Light element abundances throughout the universe provide completely independent confirmation of Big Bang nucleosynthesis. The observed ratios of hydrogen to helium, deuterium abundances in pristine gas clouds, and lithium concentrations in the oldest stars all match the precise predictions of nuclear physics calculations for the conditions that would have existed during the universe's first twenty minutes. These element abundances depend sensitively on the cosmic baryon density, allowing astronomers to determine how much ordinary matter exists in the universe through an entirely different method than CMB observations, yet yielding remarkably consistent results.

Large-scale structure observations provide a third independent confirmation of Big Bang predictions. The distribution of galaxies, galaxy clusters, and vast cosmic voids throughout the universe reflects the gravitational amplification of tiny density fluctuations that originated in the very early universe. Computer simulations of structure formation, starting from initial conditions consistent with CMB observations, successfully reproduce the observed cosmic web of matter on scales from individual galaxies to the largest surveyed volumes.

The time evolution of cosmic structure offers yet another test of Big Bang cosmology through observations of galaxies at different redshifts, corresponding to different epochs in cosmic history. Distant galaxies appear systematically different from nearby ones in ways that match theoretical predictions for how star formation, chemical enrichment, and galaxy assembly should have proceeded over billions of years. The most distant observable galaxies show properties consistent with being among the first generation of massive stellar systems to form in the universe.

Even the darkness of the night sky provides subtle evidence supporting Big Bang cosmology. In an infinite, static universe uniformly filled with stars, every line of sight should eventually intercept a stellar surface, making the entire night sky as bright as the Sun's disk. The observed darkness resolves this "Olbers' paradox" through the finite age and expansion of the universe, which limits both the number of observable stars and redshifts their light to longer wavelengths.

Precision Cosmology and Remaining Mysteries

The success of Big Bang theory has enabled the development of precision cosmology, where fundamental properties of the universe can be measured with accuracies that seemed impossible just decades ago. The combination of CMB observations, supernova distance measurements, galaxy surveys, and other techniques has established what cosmologists call the "concordance model" or Lambda-CDM cosmology, which describes the universe's composition and evolution with remarkable precision.

This standard model reveals a universe that is 13.8 billion years old, with ordinary matter comprising only about 5% of the total mass-energy content. Dark matter accounts for approximately 27% of the universe, while dark energy dominates with roughly 68% of the total. The universe appears to be geometrically flat and will likely expand forever at an accelerating rate driven by dark energy's repulsive effects.

Despite this impressive success, significant mysteries remain within the Big Bang framework. The nature of dark matter and dark energy—which together comprise 95% of the universe—remains unknown despite decades of experimental searches and theoretical investigations. The initial conditions that produced cosmic inflation, the mechanism that set up the hot Big Bang phase, require physics beyond our current understanding. The apparent fine-tuning of physical constants that allows for the formation of stars, planets, and life raises profound questions about the universe's fundamental nature.

Recent observations have also revealed some tensions within the concordance model that may indicate missing physics or systematic errors in current measurements. Different methods for measuring the Hubble constant yield slightly inconsistent results, creating the "Hubble tension" that has persisted despite intensive efforts to resolve it. Similarly, the abundance of lithium-7 observed in old stars appears significantly lower than Big Bang nucleosynthesis predictions, suggesting either gaps in our understanding of stellar physics or modifications to standard cosmological models.

The Big Bang theory has also raised philosophical questions about the nature of existence and causality that extend beyond purely scientific considerations. The apparent beginning of time itself at the Big Bang challenges traditional notions of cause and effect, while the possibility of multiple universes or eternal inflation suggests that our observable cosmos might represent only a tiny fraction of a much larger reality. These speculative ideas push the boundaries of scientific investigation while highlighting how Big Bang cosmology has fundamentally altered humanity's perspective on existence and our place in the cosmos.

The Continuing Revolution

The Big Bang theory stands today as one of science's greatest achievements, successfully describing the origin and evolution of everything we can observe in the universe. From its initial formulation based on general relativity and cosmic expansion through its current status as a precision science capable of predicting and explaining diverse phenomena across vast ranges of space and time, the theory exemplifies how human reason and careful observation can reveal nature's deepest secrets.

Yet the Big Bang's success has also revealed how much we still don't understand about the universe's fundamental nature. The discovery that ordinary matter comprises only a small fraction of cosmic mass-energy has opened entirely new frontiers in physics, while the accelerating expansion driven by dark energy suggests that the universe's future evolution may be even stranger than its dramatic past. These mysteries ensure that cosmology remains one of the most active and exciting areas of scientific research, with new observations and theoretical insights continuing to refine and extend our understanding of cosmic origins.

The story of Big Bang cosmology also demonstrates the power of the scientific method in addressing questions that might seem beyond human comprehension. Through the combination of theoretical insight, observational ingenuity, and technological innovation, scientists have reconstructed the universe's history from its first moments through the present day with remarkable detail and accuracy. This achievement represents one of humanity's greatest intellectual triumphs, revealing how physical processes operating according to comprehensible natural laws have shaped everything from the cosmic microwave background to the chemical elements that make life possible.

As we prepare to explore the universe's large-scale structure and the formation of the first stars and galaxies in subsequent chapters, the Big Bang theory provides the essential foundation for understanding how simple initial conditions evolved into the complex cosmos we inhabit today. The hot, dense, nearly uniform early universe described by Big Bang cosmology set the stage for billions of years of cosmic evolution that eventually produced stars, planets, and the emergence of consciousness capable of contemplating its own origins in the cosmic story. This remarkable journey from simplicity to complexity, from uniformity to structure, from the first light of recombination to the contemplation of distant galaxies represents perhaps the most extraordinary narrative that science has ever revealed.

Chapter 5: Dark Matter - The Universe's Hidden Scaffolding

In the grand tapestry of cosmic evolution, perhaps no mystery looms larger than the existence of dark matter—an invisible form of matter that comprises roughly 85% of all matter in the universe yet reveals itself only through its gravitational influence. This enigmatic substance neither emits, absorbs, nor reflects electromagnetic radiation, making it completely invisible to telescopes that detect light across the entire electromagnetic spectrum. Yet without dark matter's gravitational scaffolding, the universe as we know it could not exist—galaxies would never have formed, stars would never have ignited, and the cosmic web that connects the largest structures in the universe would never have taken shape.

The story of dark matter represents one of modern astrophysics' most compelling detective stories, where astronomers have used gravitational clues to infer the presence of matter they cannot see directly. Like detectives reconstructing a crime scene from circumstantial evidence, scientists have built an overwhelming case for dark matter's existence through multiple independent lines of investigation, from the motions of individual stars within galaxies to the bending of light around massive galaxy clusters billions of light-years away. This invisible universe, five times more abundant than ordinary matter, challenges our understanding of fundamental physics while driving some of the most ambitious experimental programs in science history.

The implications of dark matter's existence extend far beyond astronomy, touching the deepest questions about the nature of reality itself. If confirmed, dark matter particles represent an entirely new category of matter unknown to the Standard Model of particle physics, potentially revolutionizing our understanding of the universe's fundamental constituents. The search for these elusive particles has led to underground laboratories buried deep within mountains, space-based telescopes mapping the cosmic web, and particle accelerators probing the highest energies achievable on Earth. This chapter explores the remarkable journey from the first hints of missing matter to today's sophisticated efforts to understand the universe's dominant yet invisible component.

The Pioneer's Vision: Fritz Zwicky and the Missing Mass

The modern story of dark matter begins in 1933 with Swiss-American astronomer Fritz Zwicky, whose groundbreaking analysis of the Coma galaxy cluster revealed the first compelling evidence for matter that could not be seen. Working at the California Institute of Technology with a revolutionary new telescope at Mount Palomar Observatory, Zwicky combined his observational skills with theoretical insight to uncover one of the universe's most profound secrets. His discovery emerged from careful measurements of galaxy velocities within the Coma cluster, located roughly 320 million light-years from Earth in the constellation Coma Berenices.

Zwicky applied the virial theorem—a fundamental principle relating the kinetic and potential energies of gravitationally bound systems—to determine the cluster's total mass from the observed motions of its member galaxies. The virial theorem predicts that faster-moving galaxies require more mass to keep them gravitationally bound within the cluster, just as planets orbiting closer to the Sun must move faster to maintain stable orbits. When Zwicky measured the velocities of eight galaxies within the Coma cluster, he found velocity dispersions exceeding 2,000 kilometers per second—far higher than expected based on the cluster's visible stars and gas.

The calculation revealed a stunning discrepancy: the gravitational mass needed to explain the observed galaxy motions exceeded the visible mass by a factor of at least 400. This "missing mass" could not be accounted for by any known form of matter that emitted or reflected light. Zwicky termed this invisible component "dunkle Materie"—dark matter—launching a scientific investigation that continues today. Initially met with skepticism, Zwicky's discovery would eventually be recognized as the first evidence for what we now know is the universe's dominant form of matter.

Zwicky's prescient insight extended beyond the immediate puzzle of cluster dynamics. He recognized that this missing mass problem might be widespread throughout the universe, potentially affecting our understanding of cosmic structure and evolution on the largest scales. His work demonstrated how gravitational effects could reveal the presence of matter that remains invisible to electromagnetic observations, establishing a fundamental principle that continues guiding dark matter research today. The techniques Zwicky pioneered for measuring cluster masses through galaxy dynamics remain essential tools in modern cosmology, refined through decades of technological advancement but based on the same physical principles he first applied to the Coma cluster.

The Smoking Gun: Galaxy Rotation Curves

While Zwicky's cluster observations provided the first evidence for dark matter, the most compelling and widely recognized proof came four decades later through the painstaking work of astronomer Vera Rubin and her collaborators at the Carnegie Institution. Rubin's systematic study of galaxy rotation curves—plots showing how stellar orbital velocities change with distance from galactic centers—revealed that the dark matter problem extended far beyond galaxy clusters to individual spiral galaxies throughout the universe.

The expected behavior of galaxy rotation curves follows directly from Kepler's laws of planetary motion, which describe how orbital velocities depend on the distribution of mass within a system. In our solar system, planets farther from the Sun orbit more slowly because most of the system's mass concentrates at the center. Similarly, if galaxies contained only their visible stars and gas, stellar velocities should decrease with distance from galactic centers, following what astronomers call a "Keplerian decline". Stars in a galaxy's outer regions should orbit much more slowly than those near the center, just as Neptune orbits the Sun more slowly than Mercury.

Rubin's observations revealed something completely unexpected. Instead of declining with distance, stellar velocities in spiral galaxies remained essentially constant across their entire visible disks, creating "flat" rotation curves that defied gravitational expectations based on visible matter alone. Stars located 60,000 light-years from the Andromeda galaxy's center moved just as fast as those only 10,000 light-years from the core, despite being much farther from the galaxy's visible mass concentration. This pattern appeared consistently in galaxy after galaxy, suggesting a universal phenomenon rather than an isolated anomaly.

The flat rotation curves could be explained only if galaxies contained vast amounts of additional matter extending far beyond their visible boundaries. This dark matter would need to be distributed in roughly spherical "halos" surrounding each galaxy, with total masses five to ten times greater than all the visible stars and gas combined. Computer simulations confirmed that such dark matter halos could produce the observed flat rotation curves while remaining consistent with other aspects of galactic structure and evolution. The discovery earned widespread acceptance within the astronomical community, establishing galaxy rotation curves as perhaps the most direct and unambiguous evidence for dark matter's existence.

Modern observations have extended rotation curve studies to thousands of galaxies across a wide range of masses and morphological types, consistently confirming the need for dark matter halos. Advanced techniques using radio observations of hydrogen gas have mapped rotation curves to even greater distances from galactic centers, revealing that dark matter halos extend two to three times farther than visible galactic disks. These observations have established the basic architecture of galaxy-scale dark matter distribution while raising new questions about the fundamental nature of the particles that comprise this invisible component.

Gravitational Lensing: Seeing the Invisible

Perhaps the most elegant confirmation of dark matter comes through gravitational lensing—the bending of light by massive objects predicted by Einstein's general theory of relativity. This phenomenon provides a direct method for mapping dark matter distributions throughout the universe, independent of assumptions about stellar populations or galaxy formation processes. Gravitational lensing reveals dark matter through its fundamental property—gravitational attraction—

while remaining completely insensitive to whether that matter emits, absorbs, or reflects electromagnetic radiation.

The physics of gravitational lensing follows directly from Einstein's insight that massive objects curve spacetime, causing light rays to follow curved paths as they travel through the universe. When light from distant galaxies passes near massive foreground objects like galaxy clusters, this curvature deflects the light rays, creating distorted and magnified images of the background sources. The degree of lensing distortion depends only on the total mass of the foreground object, regardless of whether that mass consists of stars, gas, or dark matter.

Strong gravitational lensing produces some of the most spectacular images in astronomy, creating giant arcs, rings, and multiple images of background galaxies around massive galaxy clusters. The famous Einstein Cross shows four images of the same distant quasar, created when its light passes around a foreground galaxy that acts as a gravitational lens. These dramatic examples demonstrate lensing's power to reveal mass distributions that would otherwise remain hidden from electromagnetic observations.

Weak gravitational lensing provides an even more powerful tool for dark matter studies, detecting subtle distortions in the shapes of millions of background galaxies to map dark matter distributions across vast cosmic volumes. This technique measures statistical correlations in galaxy shape distortions rather than individual dramatic lensing events, allowing astronomers to construct detailed maps of dark matter structure throughout the observable universe. Large-scale lensing surveys have revealed the cosmic web of dark matter filaments predicted by computer simulations, confirming theoretical models of structure formation while providing unprecedented views of the universe's invisible scaffolding.

The Bullet Cluster provides perhaps the most compelling single piece of evidence for dark matter through gravitational lensing observations. This system formed when two massive galaxy clusters collided at speeds approaching 4,500 kilometers per second, creating a natural experiment that separated different types of matter based on their physical properties. X-ray observations show hot gas from both clusters slowed by collision forces, while optical observations reveal galaxies that passed through the collision relatively unaffected. Crucially, gravitational lensing maps show that most of the system's mass—the dark matter—remains associated with the galaxies rather than the hot gas, providing direct evidence that dark matter interacts only weakly with ordinary matter.

The Cosmic Web: Dark Matter's Architectural Role

Dark matter's most profound influence on the universe appears through its role in cosmic structure formation, providing the gravitational framework that enabled the growth of galaxies, stars, and ultimately life itself. Without dark matter's early gravitational collapse, the universe would have remained a nearly uniform sea of hydrogen and helium gas, never developing the complex hierarchical structures that characterize the modern cosmos. Computer simulations demonstrate that ordinary matter alone could not have formed galaxies within the universe's 13.8-billion-year age, making dark matter essential for explaining the existence of everything we observe.

The cosmic web represents dark matter's grandest architectural achievement—a vast network of filaments, walls, and voids that spans billions of light-years and connects the universe's largest structures. This web formed as dark matter began collapsing under its own gravity shortly after the Big Bang, creating increasingly dense regions that served as gravitational wells for ordinary matter. The intersection points of dark matter filaments became the sites where galaxy clusters eventually formed, while the filaments themselves channeled gas flows that fed star formation in developing galaxies.

Recent observations have begun revealing the cosmic web directly, confirming decades of theoretical predictions about its structure and evolution. Advanced imaging techniques can now detect the faint glow of hydrogen gas flowing along dark matter filaments, providing the first direct views of matter transport in the cosmic web. These observations show filamentary structures extending for hundreds of millions of light-years, exactly matching the predictions of computer simulations that include dark matter as the dominant mass component.

The hierarchical nature of cosmic structure formation depends critically on dark matter's properties and behavior. Large-scale computer simulations that follow billions of dark matter particles through cosmic time successfully reproduce the observed distribution of galaxies and galaxy clusters when they assume dark matter consists of "cold" particles that move slowly compared to light speed. These simulations predict specific relationships between halo masses, formation times, and internal structure that match observational data with remarkable precision. Alternative theories that modify gravity rather than invoking dark matter struggle to reproduce this detailed agreement between theory and observation.

The success of cold dark matter simulations in explaining cosmic structure provides strong indirect evidence for dark matter's particle nature, even though the fundamental identity of these particles remains unknown. The simulations reveal that dark matter halos have universal properties—their density profiles, shapes, and substructure follow predictable patterns regardless of halo mass or formation history. These universal scaling relationships emerge naturally from gravitational physics operating on collisionless dark matter particles, providing additional constraints on the properties of whatever dark matter turns out to be.

The Great Hunt: Searching for Dark Matter Particles

The overwhelming evidence for dark matter's gravitational effects has motivated some of the most sophisticated and sensitive experiments in physics, as researchers attempt to detect individual dark matter particles through their rare interactions with ordinary matter. These searches face extraordinary technical challenges because dark matter particles, by definition, interact only weakly with electromagnetic radiation and ordinary matter, making them extremely difficult to detect despite their cosmic abundance.

Underground direct detection experiments represent the most straightforward approach to dark matter discovery, using ultra-pure targets of xenon, germanium, or other materials to catch dark matter particles as they pass through Earth. These detectors must be buried deep underground and surrounded by multiple layers of shielding to block cosmic rays and other background radiation that could mimic dark matter signals. The largest current experiment, XENON1T, monitors over a ton of liquid xenon for the tiny flashes of light that would result from dark matter particle collisions. Despite years of operation, these experiments have detected no conclusive dark matter signals, placing increasingly stringent limits on how strongly dark matter can interact with ordinary matter.

Particle accelerator searches take a complementary approach, attempting to create dark matter particles through high-energy collisions between protons at facilities like the Large Hadron Collider. These experiments look for "missing energy" in collision products—energy that would be carried away by dark matter particles produced in the collisions but invisible to the detectors. Like direct detection experiments, accelerator searches have not yet discovered dark matter particles, but they provide important constraints on possible particle masses and interaction strengths.

Space-based indirect detection experiments search for the products of dark matter particle annihilation in regions where dark matter concentrates most densely. When dark matter particles collide and annihilate each other, they should produce high-energy photons, neutrinos, or other particles that can be detected by satellites like the Fermi Gamma-ray Space Telescope. These searches focus on galaxy centers, dwarf satellite galaxies, and other locations where dark matter densities reach their highest values. While several potential signals have been reported, none has yet provided definitive evidence for dark matter annihilation.

The absence of detected dark matter particles despite decades of increasingly sensitive searches has begun constraining theoretical models while opening new experimental directions. Traditional Weakly Interacting Massive Particles (WIMPs)—long considered the most promising dark matter candidates—are being excluded across wide ranges of masses and interaction strengths. This has motivated interest in alternative dark matter candidates including axions, sterile neutrinos, and more exotic possibilities that might interact with ordinary matter through different mechanisms.

The Cosmic Microwave Background: Precision Cosmology and Dark Matter

The cosmic microwave background radiation provides some of the most precise and model-independent evidence for dark matter through its detailed temperature and polarization patterns across the sky. These subtle variations, measured with exquisite precision by satellites like WMAP and Planck, preserve information about the early universe when it was only 380,000 years old, including the effects of dark matter on cosmic evolution. The CMB patterns provide independent confirmation of dark matter's existence while constraining its properties and abundance with remarkable accuracy.

The physical processes that created CMB temperature fluctuations depend sensitively on the universe's matter content and composition during the first few hundred thousand years after the Big Bang. Ordinary matter and radiation were tightly coupled during this epoch, creating acoustic oscillations—sound waves—in the primordial plasma that left characteristic signatures in the CMB temperature patterns. Dark matter, which does not interact electromagnetically, influenced these oscillations only through its gravitational effects, creating subtle but measurable modifications to the expected CMB patterns.

Detailed analysis of CMB observations reveals that dark matter comprises approximately 26.8% of the universe's total mass-energy content, while ordinary matter accounts for only 4.9%. This measurement agrees remarkably well with the dark matter abundances inferred from galaxy rotation curves, gravitational lensing, and other independent techniques, providing strong confirmation that a single dark matter component can explain diverse astronomical observations. The precision of CMB measurements has reached the level where small changes in dark matter properties would produce detectable effects, allowing increasingly stringent tests of theoretical models.

The CMB also provides crucial constraints on when and how dark matter began influencing cosmic structure formation. Computer simulations that match CMB observations require dark matter to have been "cold"—meaning the particles moved slowly compared to light speed during the early universe. Hot dark matter candidates like neutrinos would have prevented structure formation on galactic scales, while warm dark matter would have modified the CMB patterns in ways that disagree with observations. These constraints help narrow the range of possible dark matter particle properties while ruling out many alternative theories.

Recent CMB polarization measurements have opened new avenues for dark matter studies by probing even earlier epochs in cosmic history. These observations can potentially detect the gravitational wave signatures of cosmic inflation while placing additional constraints on dark matter interactions during the universe's first moments. Future CMB experiments may achieve sufficient sensitivity to detect dark matter effects that are currently below the noise level, providing even more stringent tests of theoretical models.

The Dark Universe Revealed

The evidence for dark matter now spans every scale of cosmic structure, from individual galaxies to the largest observable volumes of the universe, creating one of the most compelling cases for new physics beyond the Standard Model. Multiple independent lines of investigation—galaxy dynamics, gravitational lensing, cosmic microwave background observations, Big Bang nucleosynthesis, and computer simulations—all converge on the same conclusion: the universe contains roughly five times more dark matter than ordinary matter. This convergence provides confidence that dark matter represents a real physical phenomenon rather than a failure of our understanding of gravity or stellar astrophysics.

The success of the cold dark matter paradigm in explaining cosmic structure formation represents one of modern cosmology's greatest triumphs, successfully predicting the properties of galaxies and galaxy clusters across a wide range of scales and epochs. Computer simulations based on dark matter physics can reproduce the observed cosmic web, galaxy clustering patterns, and even detailed properties of individual dark matter halos with remarkable accuracy. This predictive power strongly supports the dark matter hypothesis while providing a framework for understanding cosmic evolution from the earliest moments to the present day.

Yet significant mysteries remain within the dark matter paradigm, ensuring that this field will continue driving fundamental research for decades to come. The complete absence of detected dark matter particles despite increasingly sensitive searches suggests that these particles may interact even more weakly with ordinary matter than originally expected, or through mechanisms not yet explored experimentally. Small-scale discrepancies between simulations and observations hint that dark matter physics may be more complex than the simplest models assume, possibly involving self-interactions or other exotic properties.

The broader implications of dark matter extend far beyond astrophysics, potentially revolutionizing our understanding of fundamental physics and the nature of reality itself. If dark matter consists of new types of particles, their discovery would represent the first detection of matter beyond the Standard Model, opening entirely new frontiers in particle physics. The dark matter problem also highlights how much we still don't know about the universe's basic constituents, reminding us that even our most successful theories may represent only incomplete glimpses of a much richer and more complex reality.

As we continue our cosmic journey through subsequent chapters exploring stellar evolution, black holes, and high-energy phenomena, dark matter's invisible presence will provide the gravitational stage upon which all these dramatic cosmic events unfold. From the nuclear furnaces burning within stellar cores to the most energetic explosions in the universe, everything we observe operates within the gravitational framework established by dark matter's cosmic architecture. Understanding this hidden scaffolding proves essential for comprehending not only how the universe evolved to its present state, but also how it will continue evolving toward an uncertain but fascinating future dominated by dark matter's persistent gravitational influence.

The story of dark matter discovery demonstrates science's power to reveal hidden aspects of reality through careful observation, theoretical insight, and technological innovation.

What began as Fritz Zwicky's puzzling observations of galaxy cluster dynamics has evolved into a comprehensive picture of cosmic structure and evolution that successfully explains phenomena across vast ranges of space and time. This ongoing investigation continues pushing the boundaries of experimental technique and theoretical understanding, ensuring that the mystery of dark matter will remain one of science's most compelling and consequential frontiers for generations to come.

Chapter 6: Dark Energy - The Universe's Accelerating Expansion

In the pantheon of scientific discoveries that have fundamentally altered humanity's understanding of the cosmos, few rival the startling revelation that emerged in the final years of the twentieth century: our universe is not merely expanding, but its expansion is accelerating. This discovery shattered long-held assumptions about cosmic evolution and introduced one of the most profound mysteries in modern science—dark energy, an enigmatic force that comprises roughly 70% of the universe yet remains almost completely unknown. The story of dark energy's discovery represents a triumph of observational astronomy, a testament to human ingenuity, and a humbling reminder of how much we still don't understand about the cosmos we inhabit.

The implications of cosmic acceleration extend far beyond academic curiosity. Dark energy's dominance over ordinary matter and even dark matter means that this mysterious component will ultimately determine the fate of everything we can observe, from the smallest galaxies to the largest cosmic structures. Understanding dark energy has become one of the defining challenges of twenty-first-century physics, driving the construction of ambitious new telescopes, inspiring novel theoretical frameworks, and potentially pointing toward revolutionary new physics that could reshape our understanding of space, time, and gravity itself.

The journey toward discovering dark energy began with the assumption that cosmic expansion, set in motion by the Big Bang, should gradually slow down under the influence of gravity. Just as a ball thrown upward eventually slows and falls back to Earth, cosmologists expected the universe's expansion to decelerate as the gravitational attraction between galaxies gradually overcame the initial outward impulse. This reasonable expectation, supported by decades of theoretical work, would prove to be spectacularly wrong, leading to one of the most surprising and consequential discoveries in the history of astronomy.

The Nobel Prize Discovery: When the Universe Surprised Us

The late 1990s witnessed an unprecedented collaboration between two independent research teams, each pursuing the same ambitious goal: measuring how the universe's expansion had changed over cosmic time. The Supernova Cosmology Project, led by Saul Perlmutter at Lawrence Berkeley National Laboratory, and the High-z Supernova Search Team, headed by Brian Schmidt and including Adam Riess, both sought to use distant exploding stars as cosmic lighthouses to map the universe's expansion history. Neither team expected their observations to revolutionize cosmology, but their meticulous work would earn them the 2011 Nobel Prize in Physics and fundamentally alter our understanding of cosmic evolution.

The key to their breakthrough lay in Type Ia supernovae—stellar explosions that occur when white dwarf stars in binary systems accumulate matter from companion stars until they reach a critical mass threshold. These cosmic catastrophes proved invaluable as "standard candles" because they all reach nearly identical peak brightness, allowing astronomers to determine their distances by comparing apparent brightness with known absolute brightness. By measuring both the distance and redshift of these supernovae, the teams could trace how the universe's expansion rate had evolved over billions of years.

The expectation was clear: if gravity had been slowing the universe's expansion, then distant supernovae—observed as they were billions of years ago—should appear brighter than expected in a universe expanding at constant rate. The reasoning was straightforward: in a decelerating universe, these ancient explosions would not have traveled as far from us, making them appear closer and brighter than in a coasting universe. The teams prepared their telescopes and analysis procedures to measure this expected brightening, anticipating they would quantify how much cosmic expansion had slowed over time.

Instead, they discovered the opposite. The distant supernovae appeared dimmer than expected, not brighter. Adam Riess, analyzing data from the High-z Supernova Search Team in autumn 1997, initially thought his calculations contained an error when they yielded what seemed like a nonsensical negative mass for the universe. Only after extensive checking did the teams realize they had discovered something far more profound: the universe's expansion was not slowing down but speeding up.

The implications were staggering. For distant supernovae to appear dimmer than expected, they must be farther away than anticipated in a matter-dominated universe. This could only occur if cosmic expansion had accelerated over the billions of years since these ancient explosions occurred, carrying the supernovae to greater distances than simple coasting expansion would predict. The discovery meant that some unknown form of energy—soon dubbed "dark energy"—was not only present throughout space but was powerful enough to overcome gravity's attractive influence and drive accelerating expansion.

Both teams announced their results in 1998, providing independent confirmation of this revolutionary finding. The discovery was immediately recognized as transformational, earning the designation "Breakthrough of the Year" from Science magazine. The subsequent quarter-century of observations has only strengthened the evidence for cosmic acceleration while revealing the profound implications of dark energy's dominance.

The Nature of Dark Energy: Theoretical Attempts to Understand the Unknown

The discovery of cosmic acceleration immediately raised fundamental questions about the nature of the force driving this unexpected behavior. Dark energy exhibits properties unlike any known form of matter or energy, requiring theoretical physicists to explore exotic possibilities that challenge our understanding of space, time, and the quantum vacuum itself. The quest to understand dark energy has opened new frontiers in both theoretical physics and observational cosmology, potentially pointing toward revolutionary discoveries about the fundamental nature of reality.

The simplest explanation for dark energy invokes Einstein's cosmological constant—a mathematical term he introduced in 1917 to allow for a static universe but later called his "biggest blunder" when cosmic expansion was discovered. The cosmological constant represents a constant energy density inherent in empty space itself, creating a repulsive pressure that remains unchanged even as the universe expands. This vacuum energy would naturally drive accelerating expansion once the universe became large and diffuse enough for dark energy's repulsive effects to overcome matter's attractive gravity.

While mathematically elegant, the cosmological constant faces a profound theoretical challenge known as the cosmological constant problem. Quantum field theory predicts that empty space should contain enormous amounts of vacuum energy from the constant creation and annihilation of virtual particles. However, these theoretical calculations yield a vacuum energy density that exceeds observational limits by approximately 120 orders of magnitude—perhaps the worst theoretical prediction in physics history. This staggering discrepancy suggests either that our understanding of quantum mechanics is fundamentally incomplete or that some unknown mechanism cancels most of the predicted vacuum energy.

Alternative explanations for dark energy include quintessence—a dynamic scalar field that can change in strength over time. Named after the classical "fifth element," quintessence fields could evolve as the universe expands, potentially explaining why dark energy's effects became noticeable only in the relatively recent cosmic past. Unlike the cosmological constant, quintessence allows for dark energy that strengthens or weakens over time, creating observational signatures that could distinguish it from constant vacuum energy.

Some quintessence models exhibit "tracker" behavior, where the field's energy density closely follows that of radiation or matter during early cosmic epochs before transitioning to drive acceleration in the recent past. This tracking property could solve the so-called "coincidence problem"—the puzzle of why dark energy began dominating cosmic evolution at roughly the same time complex structures and life emerged in the universe. Without this coincidence, dark energy might have dominated so early that galaxies and stars never formed, or so late that we wouldn't observe its effects today.

Even more exotic possibilities include phantom dark energy, where the equation of state parameter w becomes less than -1. Phantom energy possesses negative kinetic energy and could drive such rapid cosmic acceleration that it eventually tears apart all bound structures in the universe. This scenario, known as the "Big Rip," would see phantom energy's density increase as the universe expands, leading to exponentially accelerating expansion that ultimately destroys galaxies, stars, planets, and even atoms themselves.

Modified gravity theories provide yet another approach to explaining cosmic acceleration without invoking new forms of energy. These models propose that Einstein's general relativity breaks down on cosmic scales, with the apparent acceleration arising from modifications to gravitational physics rather than dark energy. Such theories must explain why gravity appears normal in the solar system and galaxy scales while producing repulsive effects on cosmological distances—a challenge that has led to sophisticated "screening mechanisms" that hide modified gravity in high-density environments.

Recent Discoveries: Is Dark Energy Evolving?

The past few years have witnessed potentially revolutionary developments in dark energy research, with multiple independent studies suggesting that dark energy's influence may be changing over cosmic time. These findings, if confirmed, would rule out the cosmological constant as a complete explanation for cosmic acceleration while opening new avenues for understanding dark energy's fundamental nature.

The Dark Energy Spectroscopic Instrument (DESI), an ambitious international collaboration involving over 900 researchers from more than 70 institutions, has been at the forefront of these discoveries. DESI creates detailed three-dimensional maps of the universe using nearly 15 million galaxies and quasars, allowing unprecedented precision in measuring how cosmic expansion has evolved over the past 11 billion years. The instrument tracks dark energy's influence through baryon acoustic oscillations—subtle patterns in galaxy clustering that serve as "standard rulers" for measuring cosmic distances and expansion rates.

DESI's results consistently show hints that dark energy may be weakening over time, with statistical significance reaching 4.2 sigma when combined with other cosmological measurements. This level of evidence, while not yet meeting the gold-standard 5-sigma threshold for a definitive discovery, represents compelling indication that the standard cosmological model may require fundamental modifications. The preference for evolving dark energy appears across multiple datasets, including cosmic microwave background observations, supernova measurements, and gravitational lensing studies.

Complementing DESI's findings, recent analysis of the largest Type Ia supernova dataset ever assembled has provided additional evidence for time-varying dark energy. Researchers from the Supernova Cosmology Project compiled observations of 2,087 supernovae from different experiments, carefully standardizing the measurements to create a homogeneous sample spanning cosmic history. This unprecedented dataset shows subtle but consistent deviations from the predictions of constant dark energy, supporting the hypothesis that dark energy's equation of state has evolved over billions of years.

The implications of evolving dark energy extend far beyond academic interest. If dark energy is indeed changing over time, this could indicate new physics beyond the Standard Model of particle physics, potentially involving scalar fields, modified gravity, or exotic quantum effects. Such discoveries might also resolve some persistent tensions in cosmology, including the "Hubble tension"—the disagreement between different methods of measuring the universe's current expansion rate.

However, the evidence for evolving dark energy remains tentative, and alternative explanations for the observed patterns cannot yet be ruled out. Systematic errors in observations, unaccounted-for astrophysical effects, or subtle biases in data analysis could potentially mimic the signatures of time-varying dark energy. The cosmological community remains appropriately cautious, emphasizing the need for independent confirmation and additional observations before declaring a revolutionary discovery.

The Cosmic Future: Dark Energy's Ultimate Consequences

Dark energy's dominance over the universe's total energy budget means that this mysterious component will ultimately determine the fate of everything we can observe. Current observations suggest that dark energy comprises approximately 68-70% of the universe's mass-energy content, dwarfing both ordinary matter (5%) and dark matter (27%). This dominance implies that dark energy's properties and evolution will shape cosmic history for trillions of years to come, long after the last stars have formed and died.

If dark energy behaves like Einstein's cosmological constant—remaining forever constant as the universe expands—

then the cosmos faces a gradual but inexorable slide toward maximum entropy known as the "Big Freeze" or "heat death". In this scenario, the accelerating expansion will eventually separate galaxies so far that they disappear beyond each other's observable horizons. Star formation will continue for roughly 100 trillion years until the supply of gas needed to fuel new stars is exhausted. Existing stars will gradually burn out, leaving behind white dwarfs, neutron stars, and black holes as the only remnants of once-glorious stellar populations.

Over vastly longer timescales, even these stellar corpses will succumb to dark energy's influence. Proton decay, if it occurs as predicted by some theories, will disintegrate atomic nuclei over timescales of 10^{34} years or more. Black holes will slowly evaporate through Hawking radiation, with stellar-mass black holes disappearing in about 10^{67} years and supermassive black holes requiring 10^{100} years to completely evaporate. The final state of the universe will be a cold, empty expanse approaching absolute zero temperature, containing only elementary particles scattered across incomprehensible distances.

However, if dark energy is evolving—as recent observations suggest might be possible—then the universe's ultimate fate could be dramatically different. Strengthening dark energy, described by phantom energy models where $w < -1$, could drive such rapid acceleration that bound structures throughout the universe are literally torn apart. This "Big Rip" scenario would begin with galaxy clusters dispersing as phantom energy overcomes their gravitational cohesion. Individual galaxies would follow, then solar systems, planets, and ultimately atoms themselves as phantom energy's repulsive effects become strong enough to overcome nuclear forces.

The timeline for a Big Rip depends on how rapidly phantom energy strengthens, but some models predict this cosmic catastrophe could occur in as little as 20-30 billion years. Unlike the gradual heat death predicted for constant dark energy, the Big Rip would represent a violent end to cosmic structure, with the fabric of spacetime itself potentially becoming unstable as phantom energy approaches infinite density.

Alternatively, weakening dark energy could eventually allow gravity to reassert dominance, potentially leading to a "Big Crunch" where cosmic expansion reverses and the universe collapses back into a hot, dense state similar to its beginning. Some theories suggest this collapse-expansion cycle could repeat indefinitely, creating an eternally oscillating universe. However, current observations provide no evidence for such behavior, and most theoretical models predict that any eventual collapse would be preceded by trillions of years of continued expansion.

The Search Continues: Current and Future Dark Energy Investigations

The profound mystery of dark energy has inspired some of the most ambitious observational programs in astronomy history, combining ground-based surveys, space missions, and theoretical investigations to probe this enigmatic component's nature. These efforts represent unprecedented international collaboration, involving thousands of researchers and billions of dollars in instrumentation designed to answer fundamental questions about cosmic acceleration and the universe's ultimate fate.

Current dark energy investigations rely primarily on four complementary observational approaches. Type Ia supernovae continue serving as precision distance indicators, with modern surveys discovering thousands of these stellar explosions across cosmic time. Baryon acoustic oscillations provide "standard rulers" for measuring cosmic expansion through the statistical analysis of galaxy clustering patterns. Weak gravitational lensing reveals dark energy's effects through subtle distortions in galaxy shapes caused by intervening matter concentrations. Observations of galaxy cluster abundances and evolution constrain dark energy by measuring how cosmic acceleration affects structure formation over billions of years.

The Dark Energy Survey, which concluded observations in 2019 after six years of mapping nearly one-eighth of the entire sky, exemplifies the scale and precision of modern dark energy research. This project catalogued hundreds of millions of galaxies and discovered thousands of supernovae, creating the most detailed maps ever produced of matter distribution throughout the universe. While consistent with the standard cosmological model, DES results hint at subtle tensions that may indicate missing physics or systematic errors requiring resolution.

Future dark energy investigations promise even greater precision and scope. NASA's Nancy Grace Roman Space Telescope, scheduled for launch in the mid-2020s, will conduct unprecedented wide-field surveys designed to discover tens of thousands of Type Ia supernovae while mapping billions of galaxies. The European Space Agency's Euclid mission will create complementary three-dimensional maps of cosmic structure, measuring both weak lensing effects and baryon acoustic oscillations with extraordinary precision. These space-based missions will work synergistically, providing independent cross-checks while probing dark energy's properties across cosmic history.

Ground-based facilities are also pushing the boundaries of dark energy research. The Vera C. Rubin Observatory, currently under construction in Chile, will conduct the Legacy Survey of Space and Time (LSST), photographing the entire southern sky every few nights for ten years. This unprecedented time-domain survey will discover millions of variable objects and transient events, including vast numbers of supernovae for precise distance measurements. The facility's rapid cadence and enormous data volume will enable entirely new approaches to studying cosmic acceleration.

Gravitational wave astronomy represents an emerging frontier for dark energy studies, offering the possibility of "standard sirens"—distance measurements to binary black hole and neutron star mergers detected by LIGO, Virgo, and future gravitational wave observatories. When these cataclysmic events produce both gravitational waves and electromagnetic counterparts, astronomers can measure both distance (from gravitational wave amplitude) and redshift (from electromagnetic spectroscopy), providing independent probes of cosmic expansion history.

The Deeper Mystery: Dark Energy and Fundamental Physics

The dark energy mystery extends far beyond cosmology, potentially pointing toward revolutionary discoveries about the fundamental nature of space, time, and quantum mechanics. If confirmed, dark energy could represent the first evidence for physics beyond the Standard Model of particle physics, opening entirely new frontiers in our understanding of reality itself. The search for dark energy's nature has already inspired novel theoretical frameworks and may ultimately require abandoning some of our most basic assumptions about how the universe works.

The cosmological constant problem—the spectacular disagreement between theoretical predictions and observations of vacuum energy—suggests that our understanding of quantum mechanics in curved spacetime remains fundamentally incomplete. Resolving this discrepancy may require new insights into how quantum fields behave in expanding space or the discovery of previously unknown symmetries that cancel most vacuum energy contributions. Such discoveries would revolutionize both particle physics and cosmology while potentially enabling new technologies based on manipulating spacetime properties.

Alternative approaches to dark energy increasingly explore modifications to Einstein's general relativity on cosmic scales. These investigations probe whether gravity itself behaves differently at enormous distances or low accelerations, potentially revealing new aspects of spacetime geometry. Modified gravity theories must explain why gravitational effects appear normal in the solar system while producing cosmic acceleration, leading to sophisticated screening mechanisms that may have broader implications for understanding gravity's quantum nature.

The question of whether dark energy evolves over time carries profound implications for fundamental physics. Time-varying dark energy could arise from scalar fields similar to those proposed in theories of cosmic inflation, potentially connecting dark energy to the earliest moments of cosmic history. Such connections might reveal deep relationships between seemingly unrelated phenomena, from the universe's initial conditions to its ultimate fate.

String theory and other approaches to quantum gravity also offer potential insights into dark energy's nature. These theoretical frameworks naturally predict the existence of numerous scalar fields and extra spatial dimensions that could manifest as dark energy under appropriate conditions. While highly speculative, such ideas demonstrate how dark energy research intersects with humanity's most fundamental questions about space, time, and the nature of existence itself.

The story of dark energy discovery illustrates both the power and limitations of human knowledge. In less than three decades, astronomers have progressed from assuming cosmic expansion must be slowing to discovering it is accelerating, fundamentally altering our understanding of the universe's fate. Yet this triumph of observation has also revealed the vast extent of our ignorance—we can describe dark energy's effects with mathematical precision but have virtually no understanding of what it actually is.

As we stand at the threshold of a new era in dark energy research, with revolutionary space missions and ground-based surveys poised to provide unprecedented precision, the mystery deepens rather than resolves. Recent hints that dark energy may be evolving over time could herald discoveries as transformative as the original detection of cosmic acceleration. Whether dark energy represents Einstein's cosmological constant, exotic scalar fields, modified gravity, or something even more exotic, its ultimate explanation will likely reshape our understanding of the cosmos and our place within it.

The search for dark energy's nature embodies humanity's greatest intellectual quest—the drive to understand our cosmic environment through careful observation, mathematical analysis, and theoretical insight. As we prepare to explore the formation and evolution of galaxies in the following chapter, dark energy's invisible presence will provide the cosmic stage upon which all stellar drama unfolds, determining whether future civilizations will inhabit a universe of endless expansion or catastrophic collapse, eternal cold or violent destruction. In seeking to understand dark energy, we confront not only the universe's deepest mystery but also the ultimate questions about the nature of existence itself.

Chapter 7: Cosmic Construction - Solar System Formation and Planetary Science

Standing beneath a star-filled sky, ancient observers could hardly have imagined that the seemingly eternal celestial dance above their heads emerged from chaos—that the orderly procession of planets, the predictable phases of the Moon, and the seasonal progression of constellations all arose from a violent, turbulent process of cosmic construction that unfolded over millions of years. The formation of our solar system represents one of nature's most remarkable architectural achievements, transforming a cold, dark cloud of gas and dust into the structured planetary system we know today, complete with rocky worlds suitable for life and giant planets that serve as cosmic guardians against wandering asteroids and comets.

Understanding how our solar system formed has profound implications that extend far beyond academic curiosity. The processes that shaped our cosmic neighborhood provide the template for understanding planetary systems throughout the galaxy, informing our search for potentially habitable worlds and helping us appreciate both the commonality and uniqueness of Earth's circumstances. The story of solar system formation also reveals how the elements forged in stellar cores became organized into the complex structures that enabled life's emergence, connecting stellar evolution to planetary science and ultimately to our own existence.

Modern studies of solar system formation benefit from an unprecedented convergence of evidence, including analysis of meteorites that preserve records of the earliest solid materials, observations of planet-forming disks around young stars, sophisticated computer simulations of gravitational interactions among thousands of objects, and the discovery of thousands of exoplanets that reveal the diversity of possible outcomes from the planet formation process. This wealth of information has transformed our understanding from simple qualitative models to detailed quantitative theories capable of making precise predictions about planetary system architecture and evolution.

Historical Foundations: The Birth of Scientific Cosmogony

The quest to understand solar system origins begins with two remarkable eighteenth-century thinkers who laid the conceptual groundwork for all subsequent theories of planetary formation. Immanuel Kant, better known for his philosophical contributions, ventured into cosmology in 1755 with his work "Universal Natural History and Theory of the Heavens," proposing that the solar system began as a cloud of dispersed particles that gradually condensed under mutual gravitational attraction. While Kant lacked the mathematical tools to develop his ideas rigorously, his central insight—that planetary systems could emerge naturally from initially chaotic conditions through the operation of physical laws—represented a revolutionary departure from earlier supernatural explanations.

Pierre-Simon Laplace, one of the greatest mathematicians of his era, provided the quantitative framework that Kant's qualitative vision required. Working four decades after Kant, Laplace developed a nebular hypothesis that began with the Sun already formed and rotating, surrounded by an extended hot atmosphere that reached beyond the orbit of the outermost planet. As this rotating solar atmosphere cooled and contracted, centrifugal forces would periodically overcome gravitational attraction, causing rings of material to separate from the central mass. These rings would subsequently condense to form the planets, naturally explaining why all planets orbit in the same direction and roughly the same plane as the Sun's rotation.

The Kant-Laplace nebular hypothesis dominated scientific thinking about solar system origins for over a century, providing an elegant explanation for the solar system's most obvious regularities. The theory successfully predicted that planets should revolve around the Sun in the same direction as solar rotation, that planetary orbits should lie nearly in the same plane, and that the spacing between planetary orbits should follow some mathematical relationship. These predictions matched observations reasonably well and established the nebular hypothesis as the first scientific theory of planetary formation.

However, the nebular hypothesis also faced significant challenges that would eventually require fundamental revisions. Most problematically, the theory predicted that the Sun should contain most of the solar system's angular momentum, since it formed first and should have retained its rapid rotation. In reality, the planets carry over 99% of the solar system's angular momentum despite containing less than 1% of its mass—a distribution that seemed impossible to reconcile with Laplace's model. Additionally, the discovery of asteroids with highly eccentric orbits, moons orbiting in retrograde directions, and other irregular features challenged the neat regularity that the classical nebular hypothesis predicted.

The Modern Solar Nebula: From Molecular Cloud to Protoplanetary Disk

Contemporary understanding of solar system formation builds upon the nebular hypothesis's fundamental insights while incorporating sophisticated knowledge of stellar formation, disk physics, and the behavior of gas and dust under extreme conditions. The modern theory begins not with an already-formed Sun, but with the gravitational collapse of a portion of a giant molecular cloud—a vast, cold region of space containing gas and dust with temperatures near absolute zero and densities thousands of times lower than the best laboratory vacuums.

The trigger for this collapse might have been a shock wave from a nearby supernova explosion, the gravitational influence of a passing star, or simply the natural tendency of the densest parts of molecular clouds to contract under their own gravity. Once begun, the collapse proceeded rapidly by astronomical standards, with the central proto-Sun and surrounding disk forming within roughly 100,000 years. During this collapse, conservation of angular momentum caused the initially slowly rotating cloud to spin increasingly rapidly, while conservation of energy led to dramatic heating as gravitational potential energy converted to thermal and kinetic energy.

The resulting protoplanetary disk exhibited a complex structure with strong temperature and density gradients that would prove crucial for determining planetary compositions and locations. Near the forming Sun, temperatures reached over 1,000 Kelvin, hot enough to vaporize most solid materials except the most refractory compounds like calcium-aluminum inclusions found in meteorites today. Further from the Sun, temperatures decreased steadily, allowing different materials to condense at different distances and creating the compositional gradient that distinguishes rocky inner planets from gas-rich outer planets.

The disk's vertical structure also showed strong gradients, with a dense, cool midplane where most solid particles concentrated and a hot, tenuous upper atmosphere heated by stellar radiation. This vertical stratification influenced both the chemistry of planet formation and the dynamics of dust settling, as particles migrated toward the midplane where they could participate in planet-building processes. The disk's evolution over time, including the gradual dissipation of gas and the radial drift of solid particles, created a complex, time-dependent environment in which planet formation proceeded through multiple distinct phases.

From Dust to Worlds: The Accretion Cascade

The transformation of microscopic dust grains into planets represents one of the most remarkable examples of hierarchical structure formation in nature, involving physical processes that span over ten orders of magnitude in size and mass scales. This accretion cascade begins with dust particles smaller than the width of human hair and culminates in objects massive enough to gravitationally bind atmospheres and reshape entire regions of the solar system.

The earliest stage involves the sticking together of dust grains through van der Waals forces, electrostatic attraction, and the formation of sintered bonds when particles collide at low velocities. This process, known as coagulation, operates most efficiently when dust particles move slowly relative to each other, allowing gentle collisions that promote sticking rather than destructive fragmentation. Computer simulations show that this grain growth can proceed rapidly under favorable conditions, building centimeter-sized aggregates within thousands of years.

However, the transition from centimeter-sized objects to kilometer-scale planetesimals—the building blocks of planets—presents what planetary scientists call the "meter-size barrier".

Objects in this size range experience strong aerodynamic drag from the surrounding gas, causing them to spiral inward toward the Sun on timescales much shorter than planet formation requires. Recent research has identified streaming instability as a promising solution to this problem, showing how the interaction between solid particles and gas turbulence can concentrate dust into dense clumps that collapse gravitationally to form planetesimals directly.

Once planetesimals form, their growth into larger bodies follows well-understood gravitational physics, though the details depend sensitively on the velocity distribution of objects in the planetary disk. Small planetesimals initially grow through a runaway process, where larger objects accumulate mass faster than smaller ones because their stronger gravity increases collision rates through gravitational focusing. This runaway growth continues until the most massive objects begin to gravitationally stir the surrounding planetesimal population, increasing collision velocities and entering what planetary scientists call the oligarchic growth phase.

During oligarchic growth, a few dozen massive protoplanets emerge in each region of the solar system, each surrounded by a feeding zone from which it accumulates most of its mass. The final stage involves collisions between these protoplanets themselves—giant impacts that can melt entire worlds and create dramatic features like Earth's Moon. Computer simulations of this late-stage accretion consistently show that terrestrial planet formation requires 50-100 million years to complete, explaining why the oldest meteorites from differentiated asteroids date to approximately 4.56 billion years ago.

The Frost Line: Architect of Planetary Diversity

Perhaps no single concept proves more important for understanding solar system architecture than the frost line—the distance from the Sun where water can condense into solid ice. This seemingly simple phase transition created a fundamental divide in the early solar system that explains why Earth and Jupiter differ so dramatically in size, composition, and structure. Beyond the frost line, the abundance of solid material available for planet building increased by factors of three to four compared to the inner solar system, where only rocky and metallic compounds could exist in solid form.

The location of the water frost line depended on both the Sun's luminosity and the opacity of the protoplanetary disk, but most models place it at approximately 3-5 astronomical units during the planet formation epoch—roughly between the current orbits of Mars and Jupiter. This placement explains several fundamental features of solar system architecture, including why the inner planets remain relatively small and rocky while the outer planets became massive enough to capture hydrogen and helium atmospheres directly from the solar nebula.

Different volatile compounds have their own frost lines at different distances, creating a series of compositional boundaries that influenced planetary formation throughout the outer solar system. Carbon monoxide, nitrogen, and other highly volatile compounds remain gaseous until much greater distances, explaining why the most distant regions of the solar system contain objects rich in these exotic ices. The existence of multiple frost lines also explains the compositional diversity observed among asteroids and comets, with objects formed at different distances retaining different mixtures of volatile and refractory materials.

Modern observations reveal that the frost line's location changed significantly over time as the young Sun's luminosity evolved and the disk's properties changed. Initially, when the Sun was still contracting and accreting material, the frost line may have been located as close as 1-2 AU, potentially allowing water-rich planetesimals to form in what is now the inner asteroid belt. As the disk evolved and the Sun reached its main sequence luminosity, the frost line moved outward, creating complex patterns of volatile distribution that influence solar system composition today.

The concept of the frost line has proven crucial for understanding exoplanetary systems, where observations reveal hot Jupiters—gas giant planets orbiting extremely close to their stars—that could not have formed in their current locations. These planets must have formed beyond their system's frost line and subsequently migrated inward, demonstrating that planetary migration represents a common outcome of planet formation processes. Such discoveries have transformed our understanding of solar system formation from a static model to a dynamic process involving significant orbital evolution.

Planetary Migration and the Late Heavy Bombardment

The solar system we observe today represents not the original architecture established during planet formation, but rather the end result of billions of years of dynamical evolution involving dramatic changes in planetary orbits. Evidence from both theoretical models and meteorite studies indicates that Jupiter and Saturn migrated substantially from their formation locations, with profound consequences for the entire solar system's structure and evolution.

The Grand Tack hypothesis provides the most widely accepted model for early giant planet migration in our solar system. According to this scenario, Jupiter initially formed beyond the current asteroid belt at approximately 3.5 AU, then migrated inward to within 1.5 AU of the Sun before Saturn's formation and subsequent migration created a gravitational resonance that reversed both planets' inward motion. This complex dance of giant planet migration explains several otherwise puzzling features of the solar system, including Mars's surprisingly small size and the asteroid belt's mixed composition.

Jupiter's inward migration would have scattered planetesimals throughout the inner solar system while depleting the region where Mars eventually formed, naturally explaining why Mars contains only about 11% of Earth's mass rather than the Earth-sized planet that simple accretion models predict. The subsequent outward migration of both giant planets scattered additional material throughout the solar system while establishing the gravitational architecture that eventually led to the Late Heavy Bombardment approximately 700 million years later.

The Late Heavy Bombardment represents one of the most dramatic events in solar system history, occurring roughly 4.1 to 3.8 billion years ago when a sudden surge of impacts cratered the Moon, Earth, and other inner planets. Lunar samples returned by Apollo missions show a clear spike in impact-melt ages during this period, providing direct evidence for this ancient catastrophe. The leading explanation involves a gravitational instability among the outer planets that scattered vast numbers of asteroids and comets into the inner solar system.

Computer simulations based on the Nice model show how gravitational interactions between Jupiter and Saturn could have destabilized the orbits of Uranus and Neptune, causing these planets to scatter planetesimals throughout the solar system as they migrated to their current positions. This late instability would have created the Kuiper Belt and Oort Cloud while sending a final wave of impactors toward the inner planets. The bombardment may have delivered significant amounts of water and organic compounds to Earth, potentially playing a crucial role in enabling the emergence of life.

Comparative Planetology: Lessons from Exoplanetary Systems

The discovery of thousands of planets orbiting other stars has revolutionized our understanding of planetary formation by revealing the full diversity of possible outcomes from the planet-building process. These exoplanetary systems display architectural features that range from familiar Earth-like configurations to exotic arrangements that challenge our theoretical understanding and expand our conception of planetary system evolution.

Perhaps most surprisingly, statistical studies of exoplanetary systems reveal that solar system-like architectures—with small rocky planets near the star and giant planets in outer orbits—represent a minority among known planetary systems. The most common exoplanets appear to be "super-Earths" and "mini-Neptunes" with sizes intermediate between Earth and Neptune, planet types that do not exist in our solar system. This discovery suggests that our solar system's architecture may be relatively unusual, raising questions about the specific conditions that led to its formation.

Hot Jupiters—gas giant planets orbiting extremely close to their host stars—provide compelling evidence that planetary migration plays a crucial role in shaping planetary system architecture. These planets complete orbits in just a few days and experience surface temperatures exceeding 1,000 Kelvin, conditions that would have prevented their formation in situ. The existence of hot Jupiters demonstrates that giant planets commonly migrate large distances from their formation locations, validating theoretical models that predict similar migration in our own solar system's past.

Multi-planet systems observed by the Kepler space telescope reveal that tightly packed planetary systems with multiple planets in resonant chains represent another common outcome of planet formation. These systems often contain five or more planets orbiting closer to their star than Mercury orbits the Sun, suggesting that inward migration and careful gravitational choreography can create stable configurations completely unlike our solar system. The prevalence of such systems indicates that our solar system's widely spaced planetary orbits may reflect the particular history of giant planet migration that characterized its formation.

The study of exoplanetary atmospheres has begun revealing the compositional diversity that results from different planet formation environments. Planets that formed closer to or farther from their system's frost line show systematically different atmospheric compositions, confirming theoretical predictions about how formation location influences planetary properties. These observations provide crucial tests of planet formation models while demonstrating how comparative studies of exoplanetary systems enhance our understanding of solar system formation processes.

Water Worlds and Habitable Zone Architecture

The search for potentially habitable exoplanets has highlighted the crucial importance of water delivery during planet formation, connecting solar system formation studies directly to astrobiology and the search for life beyond Earth. Understanding how Earth acquired its oceans has become essential for predicting which exoplanetary systems might harbor worlds suitable for life as we know it.

Multiple lines of evidence indicate that Earth's water came primarily from asteroids and comets that formed beyond the frost line and subsequently migrated into the inner solar system. Isotopic analyses of terrestrial water and meteorites suggest that carbonaceous asteroids contributed more water to Earth than comets, implying that the Late Heavy Bombardment and earlier phases of planetary migration played crucial roles in making our planet habitable. This water delivery mechanism requires specific timing—too early and the water would be lost during late giant impacts, too late and insufficient water would accumulate.

The Grand Tack model provides a natural explanation for efficient water delivery by showing how Jupiter's migration scattered water-rich asteroids from the outer solar system into Earth-crossing orbits. Without this migration, Earth might have remained a dry, Mars-like world incapable of supporting life. The model also explains why Mars contains evidence for ancient oceans despite its current arid state—the planet may have received water-rich impactors during the same migration events that hydrated Earth but lacked sufficient mass to retain its atmosphere over geological time.

Exoplanetary system architectures reveal a wide range of possibilities for water delivery and retention. Systems with hot Jupiters that migrated inward after terrestrial planet formation might have disrupted water delivery entirely, creating dry inner planets despite adequate water sources in the outer system. Conversely, systems with giant planets that remained in outer orbits might experience more efficient water delivery than occurred in our solar system, potentially creating "super-Earths" with much deeper oceans than Earth possesses.

The timing of giant planet formation relative to terrestrial planet assembly appears crucial for determining habitability outcomes. If giant planets form too early, their gravitational effects can prevent terrestrial planets from accumulating sufficient mass. If they form too late, they cannot influence water delivery during the critical phases of inner planet formation. This delicate timing may explain why Earth-like planets appear relatively rare among observed exoplanetary systems, requiring a specific sequence of events to achieve the right combination of size, composition, and water content for habitability.

The Continuing Story: Modern Solar System Evolution

Solar system formation did not end with the establishment of planetary orbits and the cessation of heavy bombardment—it continues today through more subtle but ongoing processes that slowly modify planetary properties and system architecture. Understanding these contemporary processes provides insights into the long-term evolution of planetary systems while helping predict the future development of both our solar system and exoplanetary systems observed at different evolutionary stages.

Tidal interactions between planets and their moons continue reshaping orbits over billions of years, with Earth's Moon gradually spiraling away from our planet while Mars's small moons slowly approach their host. These orbital changes affect planetary rotation rates, climate cycles, and the long-term stability of planetary spin axes, demonstrating how gravitational physics continues influencing planetary environments long after the violent early phases of system formation end.

The ongoing accretion of interplanetary dust and small asteroids provides a slow but steady source of new material to planetary surfaces, particularly noticeable on airless bodies like the Moon where micrometeorite impacts constantly modify surface composition. While this contemporary accretion rate is millions of times lower than during the Late Heavy Bombardment, it represents the modern continuation of the same basic processes that built planets from smaller components during the solar system's youth.

Recent spacecraft missions have revealed that small bodies throughout the solar system preserve detailed records of planet formation processes, providing direct observational evidence for theoretical models developed through computer simulations. The Japanese Hayabusa2 mission's analysis of samples from asteroid Ryugu confirms that this object formed in the outer solar system before migrating to its current location in the asteroid belt, validating models of planetesimal transport during giant planet migration. Similarly, NASA's OSIRIS-REx mission has gathered samples from asteroid Bennu that will provide additional constraints on early solar system conditions and processes.

The study of solar system formation has evolved into a mature science that successfully explains the major features of planetary system architecture while making testable predictions about exoplanetary systems throughout the galaxy. This success demonstrates the power of combining theoretical modeling, laboratory analysis, spacecraft observations, and astronomical surveys to understand complex astrophysical processes. As we continue discovering new exoplanetary systems and gathering more detailed information about solar system history, our understanding of planetary formation will undoubtedly continue evolving, revealing new insights into the cosmic processes that created the worlds we observe today.

The story of solar system formation connects directly to humanity's place in the cosmic order, explaining how the elements forged in stellar cores became organized into the complex structures that enabled life's emergence on at least one world. This understanding provides both humbling perspective on our cosmic insignificance and inspiring recognition of the remarkable processes that made our existence possible through billions of years of cosmic evolution. As we search for life beyond our solar system, the lessons learned from studying our own planetary system's formation provide essential guidance for identifying the most promising targets and understanding the conditions necessary for habitability throughout the galaxy.

Chapter 8: Stellar Lives and Deaths - From Main Sequence to White Dwarfs

The story of stellar evolution represents one of astronomy's greatest triumphs—a comprehensive understanding of how stars are born, live, and die that spans billions of years and connects the smallest nuclear reactions to the grandest cosmic phenomena. Every star in the night sky, from the faintest red dwarf to the most luminous blue supergiant, follows a predictable evolutionary path determined by a single fundamental property: its initial mass. This cosmic drama, playing out over timescales that dwarf human history, has shaped the chemical composition of the universe and made possible the emergence of planets, life, and ultimately our own existence.

Understanding stellar evolution provides crucial insights into the universe's past, present, and future. Stars serve as both the primary sources of energy that illuminate galaxies and the cosmic furnaces that forge the heavy elements essential for rocky planets and biological processes. The nuclear reactions burning within stellar cores follow the same physical laws that govern terrestrial laboratories, yet they operate under conditions of temperature and pressure so extreme that they transform the fundamental building blocks of matter itself. Through stellar evolution, the universe has gradually enriched itself with carbon, oxygen, silicon, iron, and dozens of other elements that began as simple hydrogen and helium in the aftermath of the Big Bang.

The predictable nature of stellar evolution allows astronomers to use stars as cosmic clocks, determining the ages of star clusters, galaxies, and even the universe itself. More massive stars evolve rapidly, burning through their nuclear fuel in millions of years, while the lowest-mass stars can shine for hundreds of billions of years—longer than the current age of the universe. This relationship between stellar mass and lifetime creates a natural laboratory for studying cosmic evolution, as astronomers can observe stars at every stage of their development within a single galaxy or star cluster.

Stellar Birth: From Molecular Clouds to Nuclear Ignition

The stellar story begins in the vast, cold expanses of molecular clouds—regions of space where temperatures hover near absolute zero and densities remain thousands of times lower than the best laboratory vacuums. These seemingly empty regions actually contain the raw materials for star formation: primarily hydrogen gas mixed with traces of heavier elements and dust grains that serve as condensation nuclei for further accumulation. Spanning dozens of light-years and containing enough material to form thousands of stars, molecular clouds represent stellar nurseries where gravity gradually overcomes thermal pressure to initiate the collapse that creates new suns.

The transition from diffuse cloud to blazing star requires overcoming several physical obstacles that initially prevent gravitational collapse. Thermal pressure from the cloud's kinetic energy opposes gravitational contraction, while magnetic fields threading through the ionized gas can provide additional support against collapse. The trigger that finally initiates star formation might come from external sources—shock waves from nearby supernova explosions, gravitational perturbations from passing stars, or simply the natural tendency for the densest regions within clouds to become gravitationally unstable once they exceed a critical mass threshold known as the Jeans mass.

Once gravitational collapse begins, the process accelerates as material falls inward under its own weight. Conservation of angular momentum causes the initially slowly rotating cloud to spin faster as it contracts, while conservation of energy converts gravitational potential energy into heat that raises the gas temperature dramatically. The collapsing material flattens into a rotating disk around a central concentration that will eventually become the star, while the increasing density and temperature in the core approach the extreme conditions necessary for nuclear fusion to begin.

The protostar phase represents a crucial transitional period when the object generates energy through gravitational contraction rather than nuclear reactions. During this phase, which can last several million years for solar-mass objects, the protostar radiates primarily in infrared light as its surface temperature remains relatively cool despite the intense heating occurring in its core. The protostar continues accreting material from the surrounding disk while its central temperature and density steadily increase toward the critical thresholds needed for hydrogen fusion—approximately 10 million Kelvin and densities exceeding 100 grams per cubic centimeter.

The moment when nuclear fusion finally ignites in the stellar core marks the true birth of a star and its entry onto the main sequence. This transition occurs when the core temperature becomes sufficient to overcome the electromagnetic repulsion between hydrogen nuclei, allowing the strong nuclear force to bind protons together in the nuclear reactions that will power the star for most of its lifetime. The onset of hydrogen burning establishes hydrostatic equilibrium—a delicate balance between the outward pressure from nuclear energy generation and the inward pull of gravity that will regulate the star's structure and evolution for millions to billions of years.

Main Sequence Life: The Nuclear Furnace

The main sequence represents the longest and most stable phase in a star's lifetime, during which hydrogen fusion in the core provides a steady energy source that maintains hydrostatic equilibrium against gravitational collapse. This phase accounts for roughly 90% of a star's total lifetime, making main sequence stars the most commonly observed objects in the galaxy and providing the foundation for understanding stellar physics and evolution. The specific nuclear reactions that power main sequence stars depend on their mass, with lower-mass stars like the Sun relying primarily on the proton-proton chain while more massive stars utilize the more temperature-sensitive CNO cycle.

The proton-proton chain operates in stellar cores with temperatures between 10-20 million Kelvin, making it the dominant energy source in stars up to about 1.3 solar masses. This remarkable sequence of nuclear reactions converts four hydrogen nuclei into one helium nucleus while releasing 26.7 MeV of energy per reaction cycle—equivalent to converting about 0.7% of the original mass into pure energy according to Einstein's famous equation $E=mc^2$. The energy release occurs through the production of positrons, neutrinos, and gamma rays, with only the neutrinos escaping directly from the stellar core while the gamma rays undergo countless scattering interactions that gradually transport energy toward the surface over timescales of thousands to millions of years.

More massive main sequence stars achieve higher core temperatures that favor the CNO cycle, where carbon, nitrogen, and oxygen nuclei serve as catalysts for converting hydrogen into helium. The CNO cycle releases approximately the same energy per reaction as the proton-proton chain but exhibits a much stronger temperature dependence, with reaction rates proportional to the 15th to 20th power of temperature rather than the fourth power characteristic of the proton-proton chain. This extreme temperature sensitivity explains why massive stars burn their fuel so much more rapidly than low-mass stars, despite containing larger hydrogen reservoirs.

The mass-luminosity relationship represents one of the most fundamental scaling laws in stellar astrophysics, connecting a star's mass directly to its energy output and lifetime. Main sequence stars exhibit a remarkably tight correlation between mass and luminosity, with luminosity roughly proportional to the 3.5 power of mass for most stellar masses. This relationship means that a star twice as massive as the Sun shines about 11 times brighter, while a star ten times more massive radiates approximately 3,000 times more energy. Since more massive stars contain only proportionally more fuel but burn it at rates that increase much faster than their mass, they exhaust their hydrogen supplies far more rapidly than their lower-mass counterparts.

The main sequence lifetime of stars ranges from over 100 billion years for the lowest-mass red dwarfs to only a few million years for the most massive blue supergiants. The Sun, with a main sequence lifetime of approximately 10 billion years, represents a middle ground in this vast range and has already consumed about half of its core hydrogen during its 4.6 billion year history. The extreme longevity of low-mass stars means that every red dwarf ever formed in the universe's 13.8 billion year history remains on the main sequence today, while massive stars complete their entire evolutionary cycle within timescales short enough for astronomers to observe multiple generations within recorded history.

Post-Main Sequence Evolution: The Red Giant Phase

When a star exhausts the hydrogen in its core, the stable balance that characterized its main sequence lifetime comes to an end, initiating a dramatic series of structural changes that transform the star from a compact, stable object into an enormous, luminous red giant. This evolutionary transition begins when the core, now composed primarily of helium "ash" from billions of years of hydrogen fusion, can no longer maintain the nuclear reactions needed to support the star's outer layers against gravitational collapse. The core begins to contract under its own weight, releasing gravitational potential energy that heats the surrounding layers and ignites hydrogen fusion in a shell around the inactive core.

The physics of shell burning differs dramatically from core burning and drives the remarkable changes that characterize red giant evolution. As the helium core contracts and heats up, it causes the hydrogen-burning shell to move outward through layers of the star that had never previously experienced nuclear fusion. This shell source proves less stable than core burning, causing the star's energy output to increase substantially while the outer envelope expands and cools in response to the increased energy flow. The result is a star that becomes hundreds of times more luminous while growing to dozens of times its original radius, creating the cool, extended atmosphere that gives red giants their characteristic color.

During the red giant phase, convection plays a crucial role in mixing processed material throughout the stellar interior, bringing helium and heavier elements created in the core to the stellar surface in events called dredge-up episodes. These mixing events alter the star's surface composition and can dramatically affect its subsequent evolution by changing the opacity and nuclear reaction rates in different layers. The first dredge-up occurs early in red giant evolution and brings helium produced during main sequence hydrogen burning to the surface, while later dredge-ups can transport carbon and other products of advanced nuclear burning to the stellar atmosphere.

The red giant phase culminates when the contracting helium core finally reaches the temperature and density necessary for helium fusion to begin through the triple-alpha process. This nuclear reaction sequence combines three helium nuclei to form carbon, requiring temperatures exceeding 100 million Kelvin due to the difficulty of bringing three particles together simultaneously. For stars similar to the Sun, the onset of helium burning occurs explosively in an event called the helium flash, when the entire degenerate core ignites almost simultaneously and releases enormous amounts of energy in a few seconds. More massive stars experience a gentler ignition of helium burning because their cores never become degenerate, allowing helium fusion to begin more gradually.

The helium-burning phase, also known as the horizontal branch for low-metallicity stars, represents a second period of relative stability in stellar evolution. During this phase, which lasts roughly 100 million years for solar-mass stars, the star maintains a more compact structure than during the red giant phase while simultaneously burning helium in the core and hydrogen in a surrounding shell. The complex nuclear physics of this phase produces carbon and oxygen in the core while continuing to generate the heavy elements that will eventually enrich the interstellar medium when the star sheds its outer layers.

Stellar Mass Loss and the Asymptotic Giant Branch

As stars evolve beyond the helium-burning phase, they enter an increasingly unstable period characterized by dramatic mass loss, complex nuclear burning, and violent structural changes known as the asymptotic giant branch (AGB) phase. This phase represents the final chapter in the evolution of low and intermediate-mass stars, during which they can lose 50-70% of their total mass through stellar winds and episodic mass ejection events that gradually strip away their outer layers to expose the hot, dense core that will become a white dwarf.

The AGB phase is characterized by thermal pulses—periodic instabilities in helium shell burning that cause dramatic changes in the star's structure and luminosity every 10,000 to 100,000 years. These thermal pulses occur because helium burning in a thin shell is inherently unstable, with small increases in temperature leading to explosive increases in nuclear reaction rates that briefly make the helium shell thousands of times more luminous than the rest of the star. Each thermal pulse creates a convective zone that mixes processed material from the core region toward the surface, potentially altering the star's chemical composition through successive third dredge-up events that can transform oxygen-rich stars into carbon-rich stars.

Stellar winds during the AGB phase are driven by a combination of pulsations, radiation pressure, and dust formation in the cool, extended stellar atmospheres. As processed material from the stellar interior reaches the surface through dredge-up events, it can condense into dust grains that efficiently absorb stellar radiation and transfer momentum to the surrounding gas through radiation pressure. This process creates mass-loss rates between 10^{-8} and 10^{-4} solar masses per year—up to 100,000 times higher than the mass loss rate of the Sun—with wind velocities typically ranging from 5 to 30 kilometers per second.

The chemical enrichment of the interstellar medium represents one of the most important consequences of AGB evolution, as these stars serve as the primary sources of carbon, nitrogen, and s-process elements like strontium, barium, and lead. The third dredge-up events that characterize thermal pulses bring carbon produced by helium burning to the stellar surface, where it can form carbon-bearing molecules and dust grains that survive ejection into space. Simultaneously, the neutron flux produced by nuclear reactions in the helium-burning shell enables slow neutron capture processes that build elements heavier than iron through successive neutron additions to seed nuclei.

The transition from AGB star to planetary nebula occurs when mass loss finally exposes the hot, dense core while a fast wind from the central star sweeps up and ionizes the previously ejected material. This process creates some of the most beautiful objects in the universe—planetary nebulae—whose diverse morphologies reflect the complex interaction between stellar winds, magnetic fields, and binary companions during the final phases of stellar evolution. The central star, now revealed as a white dwarf with a surface temperature exceeding 50,000 Kelvin, ionizes the surrounding gas to create the spectacular emission nebula that will gradually expand and fade over thousands of years.

White Dwarfs: The Final Fate

White dwarf stars represent the final evolutionary endpoint for more than 97% of all stars, including our Sun, marking the transition from active nuclear burning to a slow cooling phase that will continue for trillions of years. These remarkable objects pack roughly half the mass of the Sun into a volume the size of Earth, creating surface gravities 100,000 times stronger than Earth's while maintaining surface temperatures initially exceeding 100,000 Kelvin. Understanding white dwarf physics requires quantum mechanics on astronomical scales, as these stars are supported against gravitational collapse by electron degeneracy pressure rather than the thermal pressure that supports normal stars.

The formation process that creates white dwarfs begins during the final phases of AGB evolution, when mass loss finally removes the star's outer envelope to expose the hot carbon-oxygen core that had been hidden beneath thousands of kilometers of stellar material. This newly exposed core, with a mass typically between 0.5 and 0.8 solar masses, represents the accumulated "ash" from billions of years of hydrogen and helium burning compressed into a volume smaller than Earth. The extreme density achieved in white dwarf interiors—roughly one million times denser than water—creates conditions where normal gas pressure becomes negligible compared to the quantum mechanical pressure arising from the Pauli exclusion principle.

Electron degeneracy pressure provides the mechanism that prevents white dwarfs from collapsing further under their own gravity, creating a fundamentally different relationship between mass and radius than exists in normal stars. Unlike main sequence stars, where more massive objects are larger, white dwarfs become smaller as their mass increases because higher mass requires higher degeneracy pressure, which can only be achieved through greater compression. This inverse mass-radius relationship continues until stars approach the Chandrasekhar limit of approximately 1.44 solar masses, beyond which electron degeneracy pressure becomes insufficient to prevent gravitational collapse.

The cooling evolution of white dwarfs provides astronomers with one of the most reliable methods for determining the ages of stellar populations and constraining the history of star formation in galaxies. Newly formed white dwarfs begin their cooling phase with surface temperatures exceeding 100,000 Kelvin, but they gradually radiate away their thermal energy over timescales measured in billions of years. The cooling rate depends on the white dwarf's mass and composition, with more massive objects cooling faster initially due to their smaller surface area but retaining heat longer in their dense interiors.

Recent observations have revealed additional complexity in white dwarf cooling through the discovery of crystallization and phase separation processes occurring in their dense interiors. As white dwarfs cool below approximately 6,000 Kelvin, the carbon and oxygen nuclei in their cores begin forming crystalline structures, releasing latent heat that significantly delays the cooling process. This crystallization effect, combined with the phase separation of carbon and oxygen that creates compositional gradients within the cooling star, can extend white dwarf lifetimes by several billion years and explains observed features in white dwarf color-magnitude diagrams.

Cosmic Chemical Evolution: Stellar Alchemy and Galactic Enrichment

The death of stars represents not an ending but a transformation—the conversion of nuclear energy into heavy elements that enrich the cosmos and enable the formation of planets, moons, and ultimately life itself. Through billions of years of stellar evolution, the universe has gradually increased its metallicity from the primordial mixture of hydrogen and helium created in the Big Bang to the chemically rich environment we observe today, where elements like carbon, oxygen, silicon, and iron comprise several percent of the material in galaxies like our own Milky Way.

Different mass ranges of stars contribute distinct suites of elements to galactic chemical evolution, creating a complex pattern of nucleosynthesis that depends on stellar mass, metallicity, and evolutionary history. Low and intermediate-mass stars, those destined to become white dwarfs, primarily contribute helium, carbon, nitrogen, and s-process elements through their AGB winds and planetary nebula ejection. These stars operate as efficient element factories during their thermal pulse phases, when repeated third dredge-up events bring freshly synthesized carbon and s-process elements to their surfaces where stellar winds can carry them into the interstellar medium.

The s-process—slow neutron capture nucleosynthesis—represents one of the most important contributions of AGB stars to cosmic chemical evolution, creating approximately half of the elements heavier than iron found in the solar system. This process occurs when neutrons produced by nuclear reactions in the helium-burning shell are captured by iron-peak elements faster than they decay, building up chains of increasingly heavy isotopes that eventually reach stable configurations. Elements like strontium, barium, lanthanum, and lead owe their cosmic abundance primarily to s-process nucleosynthesis in AGB stars, making these evolved objects crucial for understanding the chemical evolution of galaxies.

The timescale for chemical enrichment by low and intermediate-mass stars extends over billions of years, creating a delayed enrichment pattern that depends on the initial mass function and star formation history of different galactic environments. Stars with masses between 1-8 solar masses require between 1-10 billion years to complete their evolution and return processed material to the interstellar medium, meaning that chemical enrichment from these objects continues throughout galactic history rather than occurring in discrete bursts. This gradual enrichment explains why the metallicity of galaxies increases steadily over cosmic time and why the relative abundances of different elements carry information about the star formation history of their host systems.

The planetary nebula phase provides the primary mechanism for dispersing the products of low and intermediate-mass stellar nucleosynthesis throughout the galaxy, creating expanding shells of gas that will eventually mix with the ambient interstellar medium. These nebulae not only return processed material to space but also contribute to the heating and ionization of the interstellar medium through the intense ultraviolet radiation from their central white dwarf stars. The morphological diversity of planetary nebulae—from simple spherical shells to complex bipolar and multipolar structures—reflects the interplay between stellar winds, magnetic fields, and binary interactions that shape the final mass loss process.

Stellar Evolution as Cosmic Clock

The predictable nature of stellar evolution transforms stars into cosmic timepieces that astronomers can use to determine ages across the universe, from individual star clusters to entire galaxies. This chronometric capability arises from the tight relationship between stellar mass and main sequence lifetime, which allows astronomers to determine when different populations of stars formed by observing which types of stars have evolved away from the main sequence. The technique of main sequence turnoff dating has become one of the most reliable methods for measuring cosmic ages and constraining the history of star formation throughout the universe.

The main sequence turnoff represents the most massive stars still undergoing hydrogen core burning in a stellar population of given age. Since more massive stars evolve faster, the turnoff mass decreases steadily with time as progressively less massive stars exhaust their core hydrogen and begin evolving toward the red giant branch. By measuring the color and magnitude of the turnoff point in star clusters, astronomers can determine the cluster's age with uncertainties of only a few percent for well-studied systems. This technique has revealed that the oldest globular clusters in our galaxy have ages of 12-13 billion years, providing important constraints on the timing of galaxy formation and the age of the universe itself.

The evolution of stellar populations also provides insights into the star formation histories of galaxies through observations of their integrated colors and spectral features. Young stellar populations appear blue due to the dominance of massive, hot stars, while older populations appear progressively redder as the most massive stars evolve away from the main sequence. The detailed spectral features of galaxies encode information about the ages, metallicities, and initial mass functions of their constituent stellar populations, allowing astronomers to reconstruct the complex histories of star formation, chemical enrichment, and structural evolution that have shaped modern galaxies.

Future observations with increasingly powerful telescopes will extend stellar chronometry to even earlier cosmic epochs, potentially observing the first generations of stars that formed when the universe contained only hydrogen and helium. These Population III stars, predicted to be much more massive than typical stars today, would have evolved rapidly and enriched the early universe with the first heavy elements produced by stellar nucleosynthesis. Finding and characterizing the remnants of these ancient stars—whether as peculiar abundances in surviving low-mass stars or as unusual white dwarfs with unique compositions—represents one of the most exciting frontiers in stellar archaeology.

Conclusion: The Stellar Legacy

The story of stellar evolution reveals the universe as a vast recycling system, where each generation of stars builds upon the nuclear achievements of its predecessors while preparing the materials for future cosmic construction projects. From the first stars that formed in the pristine hydrogen-helium environment of the early universe to the metal-rich stars forming today in galaxies enriched by billions of years of stellar evolution, this cosmic alchemy has gradually transformed the universe from simple to complex, from uniform to structured, and from sterile to potentially life-bearing.

The white dwarf stars that represent the final fate of most stellar evolution serve as both monuments to stellar achievement and repositories of cosmic history. Each white dwarf carries in its crystallizing core a record of the nuclear processes that powered its progenitor star, while its cooling history provides information about the age and formation history of its stellar population. These stellar remnants will continue cooling for trillions of years, far longer than the current age of the universe, serving as eternal witnesses to the era of stellar activity that has shaped cosmic evolution.

Understanding stellar evolution from main sequence to white dwarf provides the foundation for comprehending more exotic endpoints of stellar evolution—the neutron stars and black holes that form when more massive stars undergo catastrophic collapse rather than gentle mass loss. The boundary between these different fates, determined primarily by the initial stellar mass, creates a natural division in the stellar graveyard between the slowly cooling white dwarfs and the exotic compact objects that challenge our understanding of matter and gravity under extreme conditions.

As we prepare to explore these more extreme stellar remnants in subsequent chapters, the lessons learned from studying normal stellar evolution will prove essential for understanding how stars forge the elements that make planets and life possible while occasionally creating the most extreme objects in the universe. The stellar lives and deaths chronicled in this chapter represent just one part of the cosmic story, but they provide the foundation upon which all more complex astrophysical phenomena are built, connecting the quantum nuclear reactions in stellar cores to the grand narrative of cosmic evolution itself.

Chapter 9: Neutron Stars and Degenerate Matter - The Universe's Most Extreme Physics Laboratory

In the grand theater of cosmic evolution, few objects command as much awe and scientific fascination as neutron stars—stellar remnants so dense that a teaspoon of their material would weigh as much as Mount Everest, so compact that they squeeze more mass than our Sun into a sphere smaller than Manhattan, and so exotic that they represent natural laboratories for physics under conditions impossible to replicate anywhere else in the universe. These extraordinary objects emerge from the violent deaths of massive stars, marking the boundary between conventional stellar evolution and the realm of truly extreme physics where quantum mechanics operates on astronomical scales and matter exists in states that challenge our deepest understanding of the natural world.

Neutron stars represent the ultimate triumph of quantum physics over classical expectations, demonstrating how the Pauli exclusion principle—a purely quantum mechanical effect governing the behavior of identical particles—can support objects more massive than the Sun against the relentless pull of gravity. Their existence proves that the microscopic laws discovered in terrestrial laboratories apply even under the most extreme conditions imaginable, while simultaneously revealing new physics that emerges only when matter is compressed to densities approaching those found in atomic nuclei themselves.

The story of neutron stars connects intimately to humanity's ongoing quest to understand the fundamental nature of matter, energy, space, and time. These cosmic laboratories provide unique opportunities to study superfluidity and superconductivity on astronomical scales, to probe the behavior of matter under pressures billions of times greater than those at Earth's center, and to test Einstein's theory of general relativity in gravitational fields so intense that they warp spacetime itself into configurations barely imaginable to human intuition. Through careful observation of neutron stars, astronomers have confirmed some of physics' most profound theoretical predictions while discovering phenomena so bizarre that they expand our conception of what is possible in the natural world.

The Violent Birth: Core-Collapse Supernovae and Neutron Star Formation

The journey toward neutron star formation begins with the death throes of massive stars—objects born with at least eight to ten times the mass of our Sun that have exhausted their nuclear fuel after millions of years of stellar evolution. Unlike their lower-mass counterparts that gently shed their outer layers to become white dwarfs, these stellar giants face a catastrophic end that occurs over timescales measured not in millions of years but in mere seconds, culminating in explosions so powerful that they briefly outshine entire galaxies while forging the densest stable matter known to exist in the universe.

The final moments of a massive star's life unfold according to an inexorable physics that has been playing out throughout cosmic history. As the star burns progressively heavier elements in its core—from hydrogen to helium to carbon, oxygen, and silicon—each successive fuel source provides less energy per nuclear reaction, requiring ever-higher temperatures and densities to sustain the fusion processes that have supported the star against gravitational collapse for millions of years. This process culminates when the stellar core begins producing iron-56, the most tightly bound nucleus in nature, which cannot release energy through either fusion or fission, marking the end of the star's ability to generate the internal pressure needed to maintain equilibrium.

When fusion ceases in the iron core, catastrophe strikes with stunning rapidity. The core, now containing roughly 1.4 solar masses of iron-peak elements compressed to densities approaching those found in atomic nuclei, suddenly loses the radiation pressure that had been supporting it against its own immense gravitational pull. What follows represents one of nature's most dramatic phase transitions: the core collapses from the size of Earth to a radius of approximately 20 kilometers in less than one second, while its density increases by a factor of nearly 100 billion—a compression so extreme that it forces protons and electrons to combine into neutrons through a process called inverse beta decay.

The inverse beta decay process proves crucial for understanding how neutron stars acquire their defining characteristics. Under normal circumstances, free neutrons decay into protons, electrons, and antineutrinos with a half-life of about 15 minutes, making them unstable in ordinary matter. However, the extreme conditions during core collapse reverse this process: electrons with energies exceeding 1.8 MeV are forced to combine with protons, creating neutrons and electron neutrinos in a reaction that can only occur when matter reaches densities far beyond anything found in normal stellar interiors. This process rapidly converts most of the protons and electrons in the collapsing core into neutrons, creating the neutron-rich matter that will define the resulting stellar remnant.

The collapse continues until quantum mechanics intervenes to halt the catastrophe. When the core density reaches approximately 2-3 times nuclear saturation density—roughly 5×10^{14} grams per cubic centimeter—neutron degeneracy pressure becomes sufficient to resist further gravitational contraction. At these extreme densities, neutrons are packed so tightly together that they approach their quantum mechanical limit, creating a pressure that depends not on temperature but on the fundamental quantum properties of matter itself. This quantum pressure proves capable of supporting stellar masses against gravitational collapse, but only up to a critical limit known as the Tolman-Oppenheimer-Volkoff (TOV) mass, beyond which not even neutron degeneracy can prevent continued collapse to form a black hole.

Degenerate Matter: Quantum Mechanics on Cosmic Scales

Understanding neutron stars requires grappling with one of quantum mechanics' most counterintuitive aspects: the behavior of degenerate matter, where quantum statistical effects dominate over thermal motion and create pressures that exist even at absolute zero temperature. This exotic state of matter, first encountered in white dwarf stars but reaching its ultimate expression in neutron stars, demonstrates how the Pauli exclusion principle—Wolfgang Pauli's 1925 discovery that identical fermions cannot occupy the same quantum state—scales up from atomic physics to govern the structure of objects more massive than the Sun.

The quantum mechanical foundation of neutron star stability lies in the wave-like properties of matter described by Louis de Broglie's revolutionary insight that all particles possess wavelengths inversely related to their momentum. For neutrons confined within a neutron star, these de Broglie wavelengths become comparable to the spacing between particles, creating a regime where quantum effects dominate classical physics. The Pauli exclusion principle then requires that each neutron occupy a distinct quantum state, forcing them to fill progressively higher energy levels like students filling seats in a lecture hall where no two can occupy the same chair.

This quantum filling creates what physicists call a Fermi sea—a distribution of particles that extends up to the Fermi energy, the highest occupied energy level in the system. The pressure exerted by this degenerate neutron gas depends not on temperature, as in ordinary matter, but on the density of particles and their quantum mechanical properties. The mathematical relationship for non-relativistic neutron degeneracy pressure takes the form $P = \frac{(3\pi^2)^{2/3}\hbar^2}{5m_n}(\frac{N}{V})^{5/3}$, where the pressure increases as the 5/3 power of particle density, creating a remarkably stiff equation of state capable of supporting stellar masses.

The transition from electron degeneracy in white dwarfs to neutron degeneracy in neutron stars reflects the fundamental difference in particle masses and the quantum wavelengths they produce. Electrons, with masses roughly 1,800 times smaller than neutrons, have correspondingly longer de Broglie wavelengths and reach their quantum limit at much lower densities than neutrons. This explains why white dwarfs, supported by electron degeneracy pressure, achieve maximum masses of only about 1.4 solar masses and radii of several thousand kilometers, while neutron stars can potentially reach masses approaching 2.5 solar masses while compressing to radii of only 10-15 kilometers.

At the extreme densities achieved in neutron star cores, approaching and potentially exceeding nuclear saturation density of ρ_s=2.7×10^{14} grams per cubic centimeter, the neutron degeneracy becomes relativistic, modifying the equation of state to $P = K(\frac{N}{V})^{4/3}$, where the pressure now increases as the 4/3 power of density. This relativistic degeneracy leads to a softening of the equation of state at the highest densities, ultimately limiting the maximum mass that neutron degeneracy can support and establishing the TOV limit that divides neutron stars from black holes in the cosmic mass spectrum.

The Tolman-Oppenheimer-Volkoff Limit: The Edge of Stability

The existence of neutron stars depends critically on a fundamental limit first calculated in 1939 by Richard Tolman, J. Robert Oppenheimer, and George Volkoff—a maximum mass beyond which neutron degeneracy pressure becomes insufficient to prevent gravitational collapse to a black hole. This TOV limit represents one of the most important mass scales in astrophysics, determining whether the remnant of a massive star's death becomes a neutron star observable throughout the universe or disappears forever behind an event horizon as a black hole.

The original 1939 calculation, based on simplified assumptions about neutron star matter as an ideal gas of non-interacting neutrons, yielded a maximum mass of approximately 0.7 solar masses—a result that seemed problematically low given astronomical observations suggesting that stellar remnants with masses exceeding one solar mass should be common. This discrepancy drove decades of theoretical work aimed at understanding the complex physics of matter under extreme conditions, leading to increasingly sophisticated equations of state that incorporate nuclear interactions, relativistic effects, and exotic matter phases that might exist in neutron star cores.

Modern calculations of the TOV limit must account for the full complexity of the nuclear equation of state—the relationship between pressure, density, and energy density in matter compressed beyond nuclear saturation density. This requires solving the Tolman-Oppenheimer-Volkoff equations, the general relativistic generalization of hydrostatic equilibrium that governs neutron star structure:

$$\frac{dp}{dr} = -\frac{G\epsilon(r)M(r)}{c^2r^2}\left(1+\frac{p(r)}{\epsilon(r)}\right)\left(1+\frac{4\pi r^3 p(r)}{M(r)c^2}\right)\left(1-\frac{2GM(r)}{c^2r}\right)^{-1}$$

This equation differs significantly from Newtonian hydrostatic equilibrium by including relativistic corrections that become increasingly important as gravitational fields approach their maximum possible strength before black hole formation. The three correction factors in parentheses account respectively for relativistic relations between pressure and energy density, pressure's contribution to gravitational mass, and the curvature of spacetime itself.

Contemporary theoretical work suggests that the TOV limit lies somewhere between 2.0 and 3.0 solar masses, depending on the poorly understood behavior of matter at several times nuclear saturation density. The 2017 gravitational wave detection of merging neutron stars, event GW170817, provided observational constraints suggesting that the limit lies in the range 2.01-2.17 solar masses, while discoveries of neutron stars with precisely measured masses approaching 2.0 solar masses have pushed theoretical models toward the upper end of the predicted range. These observations demonstrate that nature has found ways to support neutron star masses significantly higher than early calculations suggested, likely through nuclear interactions and equation of state stiffening that theoretical models are only beginning to understand.

Exotic Matter: Superfluids, Superconductors, and Strange Quarks

The extreme conditions within neutron stars create environments where matter exists in phases impossible to achieve in terrestrial laboratories, potentially including superfluid neutrons, superconducting protons, and exotic quark matter phases that represent the most fundamental states of nuclear matter. These exotic phases emerge when normal thermal motion becomes negligible compared to quantum effects, allowing quantum coherence to extend throughout macroscopic volumes and creating bulk properties that differ dramatically from ordinary matter.

Neutron superfluidity represents one of the most well-established exotic phases predicted to exist in neutron star interiors, arising when neutrons form Cooper pairs analogous to those responsible for superconductivity in metals. When neutron star temperatures drop below approximately 10^9 Kelvin—a condition reached within years of formation as neutrino cooling rapidly removes thermal energy—neutrons in the dense interior can form bound pairs through nuclear interactions, creating a quantum fluid that flows without viscosity and rotates through the formation of quantized vortices. This superfluid component can decouple from the solid neutron star crust, leading to complex rotational dynamics that may explain the sudden spin-up events called pulsar glitches observed in rotating neutron stars.

Proton superconductivity likely coexists with neutron superfluidity in neutron star cores, emerging from the same BCS pairing mechanism that produces terrestrial superconductors but operating under conditions of extreme density and magnetic field strength. The superconducting protons create a quantum fluid that expels magnetic flux, confining the neutron star's enormous magnetic field to quantized flux tubes that thread through the superconductor like magnetic spaghetti. The interaction between these flux tubes and the superfluid neutron component creates complex magnetohydrodynamics that influences neutron star cooling, magnetic field evolution, and the generation of pulsar radio emission.

At the highest densities achieved in neutron star cores—potentially reaching 5-10 times nuclear saturation density—theoretical models predict that ordinary nuclear matter may undergo phase transitions to exotic states involving deconfined quarks. In normal atomic nuclei, quarks remain permanently confined within protons and neutrons by the strong nuclear force, but the extreme pressures in neutron star cores might overcome this confinement to create a quark-gluon plasma similar to conditions that existed microseconds after the Big Bang. Such quark matter could exist as a hybrid phase mixed with nuclear matter or potentially as pure quark matter throughout the inner core, fundamentally altering the neutron star's equation of state and observable properties.

Strange quark matter represents the most exotic possibility for neutron star cores, involving the conversion of ordinary up and down quarks into the heavier strange quark under extreme pressure. According to the strange matter hypothesis proposed by Edward Witten and others, strange matter might actually represent the ground state of all nuclear matter—a configuration more stable than ordinary atomic nuclei but normally inaccessible due to energy barriers that extreme density might overcome. If true, neutron star cores could contain "strangelets"—nuggets of strange matter that remain stable even if ejected from the neutron star during collisions or magnetic field reconfigurations. This possibility raises profound questions about the fundamental nature of matter and whether the atoms we observe around us represent truly stable configurations or merely metastable states that could, under the right conditions, decay into more exotic forms.

Magnetic Fields and Pulsars: Cosmic Lighthouses

Neutron stars possess magnetic fields that rank among the strongest in the universe, typically exceeding Earth's magnetic field by factors of 10^8 to 10^{12}, with the most extreme examples reaching field strengths where electromagnetic effects approach their quantum mechanical limits. These enormous magnetic fields, generated during the catastrophic collapse that creates neutron stars and potentially amplified by dynamo processes in the proto-neutron star phase, create some of the most spectacular phenomena in astronomy while providing unique laboratories for studying electromagnetism under conditions impossible to achieve on Earth.

The generation of neutron star magnetic fields likely involves multiple physical processes operating during different phases of stellar collapse and early neutron star evolution. The simplest mechanism involves magnetic flux conservation during core collapse: as the stellar core shrinks from Earth-size to city-size dimensions, magnetic field lines become compressed and concentrated, amplifying modest initial fields by factors of 10^{10} or more through geometric effects alone. However, this mechanism struggles to explain the enormous range of magnetic field strengths observed among neutron stars, from 10^{11} gauss in typical radio pulsars to over 10^{15} gauss in magnetars.

Recent theoretical work suggests that convective dynamos in proto-neutron star cores may play a crucial role in determining final magnetic field strengths. During the first 10-20 seconds after neutron star formation, vigorous convection driven by neutrino cooling and deleptonization can create complex fluid motions in the highly conducting neutron star interior. When combined with rotation, these convective flows can operate as dynamos that either amplify or suppress magnetic fields depending on the rotation rate and convection strength. Fast-rotating proto-neutron stars tend to produce weaker, multipolar magnetic fields that evolve into normal radio pulsars, while slower rotation favors stronger, dipolar fields that create magnetars.

Pulsar emission represents the most common manifestation of neutron star magnetic fields, occurring when rotating neutron stars accelerate charged particles to relativistic speeds along magnetic field lines and produce beams of coherent radio emission that sweep across space like lighthouse beams. The basic pulsar mechanism involves the strong electric fields created when a magnetic dipole rotates in vacuum, accelerating particles from the neutron star surface and creating cascades of electron-positron pairs that emit coherent radio waves through complex plasma processes not yet fully understood. The remarkable regularity of pulsar timing—with periods stable to better than one part in 10^{15} for the most precise examples—makes these objects among the most accurate clocks in the universe and enables their use as precision tests of fundamental physics.

Pulsar timing observations have provided some of the most stringent tests of Einstein's general relativity, confirming predictions about gravitational time dilation, Shapiro delay effects, and orbital energy loss through gravitational wave emission. The famous binary pulsar PSR 1913+16, discovered by Russell Hulse and Joseph Taylor, showed orbital decay precisely matching general relativity's prediction for gravitational wave energy loss, providing the first indirect evidence for gravitational waves and earning its discoverers the 1993 Nobel Prize in Physics. More recent pulsar timing arrays, involving synchronized observations of many millisecond pulsars, have achieved sufficient sensitivity to detect the stochastic gravitational wave background produced by supermassive black hole binaries throughout the universe.

Magnetars: The Universe's Strongest Magnets

At the extreme end of neutron star magnetic field distribution lie magnetars—neutron stars with magnetic fields so intense that they dominate all other forms of energy in the stellar environment and create phenomena that challenge our understanding of matter and radiation under extreme conditions. With magnetic field strengths reaching 10^{14} to 10^{15} gauss—roughly 10^{11} times stronger than the most powerful magnets created on Earth—magnetars exist in a regime where quantum electrodynamics effects become important and where the magnetic field energy density exceeds the rest mass energy density of ordinary matter.

The extraordinary magnetic fields of magnetars create conditions where the vacuum itself becomes polarized and birefringent, where photons can split into multiple particles or merge together, and where atoms deform into elongated cylinders thinner than their normal quantum mechanical wavelengths. At field strengths exceeding 4.4×10^{13} gauss—the so-called quantum critical field—photon splitting becomes possible, where single high-energy photons can spontaneously divide into electron-positron pairs, fundamentally altering the electromagnetic properties of the surrounding space. These effects create conditions where normal physics breaks down and quantum field theory predictions become directly observable on astronomical scales.

The most spectacular manifestations of magnetar physics occur during giant flares—catastrophic magnetic field reconfigurations that release more energy in a few tenths of a second than the Sun emits in 250,000 years. The August 27, 1998 giant flare from SGR 1900+14 represents the best-studied example of these phenomena, detected by multiple spacecraft and ground-based instruments as it briefly saturated gamma-ray detectors across the solar system. The flare began with an initial spike lasting only 0.2 seconds but containing roughly 10^{44} ergs of energy, followed by a gradually fading tail of pulsed emission that continued for several minutes with the 5.16-second rotation period of the neutron star.

The mechanism driving giant flares likely involves sudden reconnection events in the magnetar's twisted magnetic field, analogous to but far more powerful than solar flares. The extreme field strengths in magnetars can store enormous amounts of magnetic energy—potentially 10^{47} ergs or more—in non-potential field configurations that become unstable when the magnetic stress exceeds the neutron star crust's breaking strength. When this critical threshold is exceeded, the magnetic field undergoes catastrophic rearrangement, converting stored magnetic energy into thermal energy, particle acceleration, and electromagnetic radiation with unprecedented efficiency.

Starquakes: Cracking the Cosmic Crystal

The solid crust of neutron stars represents one of the strongest materials in the universe, capable of withstanding stresses billions of times greater than the most robust terrestrial crystals can endure. This extraordinary strength emerges from the extreme gravitational pressure that compresses the crystalline lattice of atomic nuclei and electrons into configurations where quantum mechanical effects dominate the mechanical properties. Yet even these ultra-strong materials have limits, and when those limits are exceeded, neutron stars experience starquakes—sudden fractures that can reshape the stellar surface while releasing seismic energies that dwarf the most powerful earthquakes ever recorded on Earth.

The physics underlying starquake formation involves the delicate balance between magnetic stresses and crustal strength in rapidly spinning, highly magnetized neutron stars. As magnetars evolve, their magnetic fields gradually decay through ohmic dissipation and magnetic reconnection events, causing the field geometry to change over timescales of thousands to millions of years. These changes create time-varying stresses in the neutron star crust, which becomes twisted and deformed as it attempts to follow the magnetic field evolution while maintaining mechanical equilibrium. The crust can accommodate modest deformations elastically, storing energy like a compressed spring, but when the accumulated stress exceeds the breaking strength of crystalline nuclear matter, catastrophic failure occurs.

Computer simulations of neutron star crust mechanics suggest that the breaking stress reaches approximately 10^{10} dynes per square centimeter—roughly the weight of a mountain applied to every square centimeter of surface. This enormous strength allows neutron star crusts to support "mountains" several centimeters high—deformations that would be completely negligible on Earth but represent significant asymmetries for objects only 20 kilometers in diameter. When starquakes occur, they can instantly restructure these surface features while propagating seismic waves throughout the neutron star interior at velocities approaching the speed of sound in nuclear matter.

The connection between starquakes and observable phenomena emerged through detailed analysis of magnetar giant flares, particularly the remarkable discovery of quasi-periodic oscillations in the fading tails of these events. The August 27, 1998 flare from SGR 1900+14 exhibited subtle but measurable oscillations at frequencies of approximately 28, 54, and 84 Hz, interpreted as fundamental and harmonic modes of neutron star seismic waves excited by the initial starquake. These oscillations provide direct observational evidence that giant flares involve crustal fractures that set the entire neutron star "ringing like a bell" with characteristic frequencies determined by the star's mass, radius, and internal structure.

The statistical properties of soft gamma repeater bursts—the smaller-scale cousins of giant flares—exhibit patterns consistent with self-organized criticality, suggesting that magnetar crusts exist in a state of marginal stability where small perturbations can trigger avalanches of fractures spanning many orders of magnitude in energy. This behavior resembles earthquake statistics on Earth, where the Gutenberg-Richter law describes the relationship between earthquake magnitude and frequency, but with the crucial difference that magnetar starquakes operate under gravitational stresses billions of times stronger than terrestrial tectonic forces.

Recent theoretical work has explored how starquakes might contribute to the formation of asymmetric mass distributions—the "mountains" that could make rapidly spinning neutron stars sources of continuous gravitational waves. When crustal fractures occur, they can permanently deform the neutron star surface, creating or modifying the quadrupole moment that couples to gravitational wave emission. While individual starquakes typically create deformations too small to produce detectable gravitational wave signals, the cumulative effect of many fractures over time could build mountains large enough for detection by current or future gravitational wave observatories.

The extreme conditions during starquakes also create natural laboratories for studying matter under pressures and densities impossible to achieve in terrestrial experiments. The sudden release of stored elastic energy during crustal fractures creates shock waves that briefly compress nuclear matter to several times normal nuclear density, potentially triggering phase transitions to exotic states like quark-gluon plasma. These transient conditions, lasting only microseconds, might explain some of the unusual spectral features observed in magnetar bursts while providing insights into the equation of state of ultra-dense matter.

Neutron Star Mergers and Gravitational Waves: Cosmic Collisions

The detection of gravitational waves from merging neutron stars represents one of the most revolutionary discoveries in modern astronomy, opening an entirely new window for studying the universe while confirming Einstein's century-old predictions about the fundamental nature of spacetime. The August 17, 2017 detection of event GW170817 by the LIGO-Virgo collaboration marked the beginning of multi-messenger astronomy, demonstrating how gravitational wave observations can be combined with electromagnetic counterparts to extract unprecedented information about cosmic catastrophes and the extreme physics governing them.

The gravitational wave signature of merging neutron stars differs dramatically from the black hole collisions that dominated early LIGO discoveries, providing unique insights into the properties of ultra-dense matter and the final moments of stellar evolution. Unlike black hole mergers, which produce gravitational waves primarily during the final few orbits and collision, neutron star mergers generate detectable signals for minutes or hours before coalescence as the two objects spiral inward through a complex dance governed by general relativity, tidal deformation, and the exotic physics of neutron star interiors.

The pre-merger phase of neutron star collisions creates conditions where spacetime curvature becomes so extreme that it influences the internal structure of the neutron stars themselves. As the two objects approach their final orbital separation of roughly 50 kilometers—still several neutron star radii apart—tidal forces begin deforming their shapes from perfect spheres into elongated ellipsoids aligned with the orbital motion. These tidal deformations depend sensitively on the neutron star equation of state: softer equations of state allow more deformation and stronger gravitational wave emission, while stiffer equations of state resist tidal forces and produce weaker pre-merger signals.

The gravitational wave observations of GW170817 provided the first direct constraints on neutron star tidal deformability, revealing that neutron star matter is neither extremely soft nor extremely stiff but occupies an intermediate regime consistent with nuclear theory predictions. This measurement helped resolve long-standing uncertainties about the behavior of matter at several times nuclear saturation density while providing crucial input for understanding neutron star maximum masses, radii, and the phase transitions that might occur in their cores.

The merger itself creates one of the most energetically violent events in the universe, combining the mass-energy of nearly three Suns within a volume smaller than a major metropolitan area. The collision ejects roughly 1-5% of the total neutron star mass at velocities reaching 10-30% the speed of light, creating expanding shells of neutron-rich material that will later undergo rapid neutron capture nucleosynthesis to form heavy elements. The precise amount and velocity distribution of this ejecta depend on the neutron star masses, equation of state, and the dynamics of the collision itself.

Most neutron star mergers likely produce black holes as their final outcome, since the combined mass typically exceeds the maximum mass that neutron degeneracy pressure can support. However, the formation of this black hole may be delayed by differential rotation, magnetic fields, and thermal pressure that temporarily stabilize the merger remnant as a hypermassive neutron star. This transient object, surviving for milliseconds to seconds before final collapse, represents the most extreme state of matter accessible to astronomical observation—hotter and denser than anything else in the contemporary universe.

The electromagnetic counterpart to GW170817—designated AT2017gfo—provided visual confirmation of theoretical predictions about neutron star merger outcomes while revealing the synthesis of heavy elements in real time. This "kilonova" began as a bright blue transient visible in optical telescopes, then evolved over several days into a redder, infrared-dominated source as newly synthesized heavy elements absorbed and reprocessed the emission. The total energy radiated electromagnetically approached 10^{42} ergs, powered entirely by the radioactive decay of unstable isotopes created during r-process nucleosynthesis in the merger ejecta.

The Cosmic Forge: R-Process Nucleosynthesis and Heavy Element Creation

Neutron star mergers serve as the universe's most efficient factories for creating the heaviest elements, operating through the rapid neutron capture process (r-process) that builds atomic nuclei far more massive than iron through successive neutron additions faster than radioactive decay can disrupt the growing chains. This cosmic alchemy transforms the neutron-rich matter ejected during stellar collisions into gold, platinum, uranium, and dozens of other heavy elements that comprise roughly half of all elements heavier than iron found throughout the universe today.

The r-process requires extreme conditions that exist nowhere else in the contemporary cosmos: neutron densities exceeding 10^{24} particles per cubic centimeter, temperatures of billions of degrees, and timescales measured in seconds during which nuclear reactions proceed faster than stellar evolution. These conditions emerge naturally in the expanding fireball of neutron star merger ejecta, where the decompression of neutron star matter creates a transient environment rich in free neutrons while maintaining temperatures high enough to drive rapid nuclear reactions.

The nucleosynthesis process begins when seed nuclei—typically iron-peak elements ejected during the merger—encounter the flood of neutrons released by decompressing neutron star material. In the extreme neutron flux characteristic of merger environments, these seeds can capture neutrons much faster than they undergo beta decay, driving the nuclear composition along paths far from the valley of beta stability toward increasingly neutron-rich isotopes. This process continues until the growing nuclei become so neutron-heavy that additional neutron capture becomes energetically impossible, establishing temporary equilibria at so-called "waiting points" where neutron capture and photodissociation balance.

The r-process path extends to the neutron drip line—the boundary where nuclei become so neutron-rich that additional neutrons are unbound and immediately evaporate. At these extreme neutron-to-proton ratios, nuclear physics enters largely uncharted territory where theoretical models must extrapolate far beyond the stability valley where experimental data exist. The final abundance pattern depends critically on nuclear masses, neutron capture cross-sections, and beta decay rates for isotopes so unstable that they have never been produced in terrestrial laboratories and may not exist for more than milliseconds.

The discovery of strontium in the kilonova AT2017gfo provided the first direct observational confirmation that neutron star mergers operate as r-process sites. Spectroscopic analysis of the kilonova's optical emission revealed the characteristic absorption lines of strontium—a moderately heavy element with 38 protons that represents clear evidence for neutron capture processes beyond what normal stellar nucleosynthesis can achieve. The amount of strontium detected corresponds to roughly five Earth masses, demonstrating the enormous efficiency of r-process element production during these cosmic catastrophes.

The broader implications of r-process nucleosynthesis in neutron star mergers extend far beyond academic curiosity about nuclear physics. The heavy elements produced in these events include not only precious metals like gold and platinum but also radioactive isotopes like thorium and uranium that serve as natural clocks for measuring cosmic ages. The abundance patterns of r-process elements in ancient stars provide fossil records of early nucleosynthesis, revealing how the frequency and characteristics of neutron star mergers have over cosmic time while constraining models of galactic chemical evolution and the sites responsible for producing the heaviest elements in the universe.

The r-process nucleosynthesis mechanism represents one of nature's most efficient element factories, capable of building atomic nuclei containing over 200 protons and neutrons within seconds of a neutron star merger. This remarkable process begins when the neutron-rich ejecta from colliding neutron stars expands and cools, creating conditions where free neutrons vastly outnumber the seed nuclei—typically iron-peak elements—that serve as starting points for rapid neutron capture. The neutron flux during these events reaches extraordinary densities exceeding 10^{24} neutrons per cubic centimeter, creating an environment where neutron capture proceeds much faster than the beta decay processes that would normally limit nuclear growth.

The competition between neutron capture and radioactive decay determines which elements ultimately form during r-process nucleosynthesis, with the most neutron-rich environments producing the heaviest elements including actinides like thorium and uranium. As neutrons flood into seed nuclei faster than beta decay can convert them to protons, the growing nuclei become increasingly neutron-heavy and unstable, eventually reaching configurations so exotic that they exist nowhere else in the contemporary universe. These extremely neutron-rich isotopes undergo rapid neutron capture until they reach the neutron drip line—the boundary where additional neutrons become unbound—before undergoing cascades of beta decays that convert excess neutrons to protons and establish the final elemental abundances.

The neutron star merger GW170817 and its associated kilonova AT2017gfo provided the first direct observational confirmation that these cosmic collisions operate as r-process sites, revolutionizing our understanding of heavy element production throughout the universe. Spectroscopic analysis of the kilonova's evolving light revealed clear signatures of freshly synthesized heavy elements, most notably the unambiguous detection of strontium absorption lines that could only result from rapid neutron capture processes. The presence of strontium—element 38 with a complex electron configuration that produces distinctive spectral features—provided smoking-gun evidence for r-process nucleosynthesis occurring in real time within the expanding merger ejecta.

The total mass of heavy elements produced during GW170817 reached approximately 16,000 Earth masses, including several Earth masses of gold and platinum alone, demonstrating the extraordinary efficiency of neutron star mergers as cosmic element factories. This enormous yield results from the unique conditions created when neutron star matter decompresses from nuclear density to normal stellar densities over timescales of milliseconds, maintaining neutron-to-proton ratios far exceeding those found in any other astrophysical environment. The radioactive decay of these newly synthesized heavy elements powered the kilonova's luminosity for weeks, with different isotopes contributing to the light curve's evolution as their decay products thermalized within the expanding ejecta.

The r-process abundance patterns produced in neutron star mergers match remarkably well with the heavy element compositions observed in ancient metal-poor stars, providing crucial evidence that these mergers have been operating as primary r-process sites throughout cosmic history. Stars with metallicities as low as one-thousandth of solar values exhibit r-process element ratios nearly identical to those predicted by neutron star merger models, suggesting that a small number of these rare events can enrich large volumes of the galaxy with heavy elements. This agreement between theoretical predictions and observational constraints represents a triumph of multi-messenger astronomy, connecting nuclear physics calculations of exotic neutron-rich nuclei to the chemical evolution of entire galaxies.

The delayed enrichment timescale characteristic of neutron star mergers has profound implications for understanding when heavy elements first became available for planet formation and potentially life itself. Unlike massive star supernovae that explode within millions of years and continuously enrich their surroundings, neutron star mergers require the evolution of binary systems over hundreds of millions to billions of years before coalescence occurs. This delay means that the first generations of stars and galaxies formed in a universe largely devoid of elements heavier than iron, with r-process enrichment beginning only after sufficient time had elapsed for neutron star binaries to merge.

The stochastic nature of neutron star mergers creates highly inhomogeneous distributions of r-process elements throughout galaxies, leading to dramatic variations in heavy element abundances between different stellar populations and galactic regions. Individual merger events can enrich volumes exceeding 1000 cubic parsecs with heavy elements, but the rarity of these events means that many regions may experience little r-process enrichment for billions of years. This patchy enrichment pattern explains the large scatter observed in r-process element abundances among stars of similar age and metallicity, reflecting their different exposure histories to neutron star merger events.

Future gravitational wave observations promise to revolutionize our understanding of r-process nucleosynthesis by providing statistical samples of neutron star mergers across cosmic time, enabling detailed studies of how heavy element production has evolved throughout the universe's history.

Advanced LIGO and Virgo detectors are expected to observe tens to hundreds of neutron star mergers per year once they reach design sensitivity, creating opportunities to study the diversity of merger outcomes and their associated nucleosynthesis yields. These observations will constrain the fraction of mergers that successfully launch relativistic jets, the range of ejecta masses and compositions, and the dependence of r-process production on neutron star masses and equations of state.

The connection between neutron star mergers and r-process nucleosynthesis also provides a powerful probe of fundamental nuclear physics in regimes inaccessible to terrestrial experiments. The abundance patterns produced during the r-process depend sensitively on nuclear masses, neutron capture cross-sections, and beta decay rates for extremely neutron-rich isotopes that exist only briefly during merger events. Comparing theoretical r-process calculations with observed abundance patterns therefore constrains nuclear physics models and guides experimental programs aimed at measuring properties of unstable nuclei. Future radioactive ion beam facilities may eventually produce some of the neutron-rich nuclei involved in r-process nucleosynthesis, providing direct measurements of nuclear properties crucial for understanding heavy element production.

The study of r-process nucleosynthesis in neutron star mergers represents a remarkable convergence of nuclear physics, stellar astrophysics, gravitational wave astronomy, and cosmology, demonstrating how the most extreme objects in the universe connect to the fundamental question of cosmic chemical evolution. Through these cosmic collisions, neutron stars serve not only as natural laboratories for studying matter under impossible terrestrial conditions but also as the primary architects of heavy element abundance throughout the universe. The gold in our jewelry, the platinum in catalytic converters, and the uranium in nuclear reactors all likely originated in the violent merger of neutron stars billions of years ago, connecting the most exotic physics in the cosmos to the elements that make modern technology and complex chemistry possible.

As we continue to detect gravitational waves from merging neutron stars and observe their electromagnetic counterparts across the spectrum, we are witnessing the ongoing process of cosmic element creation that has operated throughout the universe's history. These observations transform neutron star physics from a purely theoretical endeavor into an empirical science with direct connections to the chemical evolution of galaxies, the formation of planetary systems, and ultimately the emergence of the complex chemistry that enables life itself. The universe's most extreme physics laboratory has thus revealed itself as the cosmic forge responsible for creating much of the periodic table, demonstrating once again how the study of the most exotic objects in nature illuminates fundamental processes that shape our cosmic environment.

Chapter 10: Black Holes - Where Physics Meets Its Limits

The cosmos presents us with objects so extreme that they challenge the very foundations of physics, regions where space and time become so distorted that our everyday understanding of reality breaks down. Black holes represent the ultimate triumph of gravity over all other forces in the universe, creating conditions so extreme that nothing—not even light itself—can escape once it crosses a critical boundary known as the event horizon. These enigmatic objects serve as cosmic laboratories where general relativity meets quantum mechanics, where information theory confronts thermodynamics, and where the fundamental structure of spacetime reaches its most dramatic expression. Far from being merely cosmic vacuum cleaners that devour everything in their vicinity, black holes are dynamic objects that power some of the universe's most spectacular phenomena, from the brilliant jets of active galactic nuclei to the X-ray emissions of stellar binary systems. Recent breakthrough observations have finally allowed us to peer directly into the shadow cast by these invisible giants, confirming decades of theoretical predictions while opening entirely new avenues for understanding the extreme physics that governs our universe.

Historical Development - From Dark Stars to Event Horizons

The concept of objects so massive that light cannot escape their gravitational pull predates Einstein's general relativity by more than a century. In 1783, John Michell, an English country parson and brilliant natural philosopher, first proposed what we would now recognize as black holes. Working within the framework of Newton's theories and accepting the prevailing view that light consisted of particles, Michell reasoned that if light had mass, it would be subject to gravitational forces just like any other object. He calculated that a star more than 500 times the diameter of the Sun, assuming the same density, would have an escape velocity exceeding the speed of light. Such an object would be invisible, as Michell wrote, since "their light could not arrive at us." His prescient insight even suggested that these dark stars might be detectable through their gravitational effects on visible companions, a prediction that would prove remarkably accurate two centuries later.

Michell's ideas made little impression on his contemporaries, and the concept of dark stars was largely forgotten until Pierre-Simon Laplace independently proposed similar ideas in his 1796 book "Exposition du Système du Monde". The notion remained dormant through the 19th century, as physicists gradually discovered that light behaves as a wave rather than as particles with mass, seemingly invalidating the original reasoning behind dark stars.

The true theoretical foundation for black holes emerged from Albert Einstein's general theory of relativity, published in 1915. Einstein's revolutionary insight showed that gravity is not a force in the conventional sense, but rather a manifestation of curved spacetime itself. Massive objects warp the fabric of space and time, and this curvature determines how matter and energy move through the universe. While Einstein had created the mathematical framework, finding exact solutions to his field equations seemed dauntingly difficult.

Remarkably, the first exact solution came within months of Einstein's publication, discovered by Karl Schwarzschild, a German astrophysicist serving as an artillery officer on the Russian front during World War I. Working in his spare time between calculating artillery trajectories, Schwarzschild found a solution describing the spacetime around a spherically symmetric, non-rotating mass. This solution, now known as the Schwarzschild metric, revealed something extraordinary: if any mass were compressed within a critical radius—later called the Schwarzschild radius—the curvature of spacetime would become so extreme that nothing could escape from within that boundary.

Schwarzschild's wartime achievement was both brilliant and tragic. He completed his work under freezing conditions while serving his country, writing to Einstein that "The war treated me kind enough, in spite of the heavy gunfire, to allow me to get away from it all and take this walk in the land of your ideas". Unfortunately, Schwarzschild developed an autoimmune disease and died in 1916, just two months after returning from the front, never knowing how profoundly his solution would impact our understanding of the universe.

The Event Horizon - A Point of No Return

The Schwarzschild solution revealed the existence of a critical boundary surrounding any sufficiently compact mass, a spherical surface at the Schwarzschild radius where the geometry of spacetime undergoes a fundamental transition. This boundary, which we now call the event horizon, represents not a physical surface that could be touched or seen, but rather a geometric feature of spacetime itself—a point of no return beyond which escape becomes impossible regardless of the energy available.

The Schwarzschild radius can be calculated using the formula $r_s=2GM/c^2$ where G is the gravitational constant, M is the mass of the object, and c is the speed of light. For our Sun, this radius would be approximately 3 kilometers, meaning that if all the Sun's mass were compressed into a sphere with a 3-

kilometer radius, it would become a black hole. The actual radius of the Sun is about 696,000 kilometers, so it would need to be compressed by a factor of more than 200,000 to reach this critical density.

What makes the event horizon so remarkable is not just its existence, but the profound way it reveals the relative nature of space and time. An observer falling toward a black hole would experience nothing unusual as they cross the event horizon—their local clocks would tick normally, and they would feel no special sensation at the moment of crossing. However, an outside observer watching this fall would witness something entirely different. Due to gravitational time dilation, the falling observer would appear to slow down dramatically as they approach the event horizon, their image becoming increasingly redshifted and dimmer until they fade from view.

This apparent paradox arises from the fundamental relativity of time itself. Near the event horizon, spacetime curvature becomes so extreme that time dilation approaches infinity as viewed from the outside. The outside observer would never actually see the falling object cross the horizon—it would appear frozen at the boundary, gradually fading away as its light becomes infinitely redshifted. Meanwhile, from the perspective of the falling observer, they cross the horizon in a finite amount of their own time, continuing their journey toward the central singularity where the laws of physics as we know them break down entirely.

This duality of perspectives illustrates one of the most profound aspects of black holes: they force us to abandon absolute notions of space and time in favor of observer-dependent descriptions of reality. The event horizon exists as a real boundary in spacetime, but its significance depends entirely on which reference frame we choose to adopt.

Types of Black Holes - From Stellar Remnants to Galactic Giants

The universe produces black holes across an enormous range of masses, from objects just a few times more massive than our Sun to monsters containing billions of solar masses. This diversity reflects the various formation mechanisms that can create these extreme objects throughout cosmic history.

Stellar-mass black holes represent the most common type, formed through the gravitational collapse of massive stars at the end of their lives. When a star with an initial mass greater than about 20-25 times the Sun exhausts its nuclear fuel, the core can no longer generate sufficient pressure to support itself against its own gravity. If the core's mass exceeds the Tolman-Oppenheimer-Volkoff limit of approximately 3 solar masses—the maximum mass that neutron degeneracy pressure can support—nothing can prevent complete gravitational collapse. The result is a stellar-mass black hole, typically ranging from about 3 to 50 solar masses.

These stellar-mass black holes are surprisingly difficult to detect directly. In our galaxy alone, astronomers estimate there may be 100 million such objects, but fewer than 100 have been confirmed or strongly suspected. The challenge lies in their nature—they emit no light of their own and can only be detected through their gravitational effects on nearby matter or companion stars.

The most successful method for finding stellar-mass black holes involves studying X-ray binary systems, where a black hole strips material from a companion star. As this material spirals inward, forming an accretion disk, it heats to millions of degrees and radiates intensely in X-rays before disappearing beyond the event horizon. The irregular variability of these X-ray emissions, in contrast to the regular pulses from neutron star systems, often provides the first clue to a black hole's presence.

At the opposite extreme lie supermassive black holes, containing millions to billions of solar masses. These giants reside at the centers of galaxies, including our own Milky Way, where Sagittarius A* contains approximately 4.3 million solar masses compressed within a region smaller than our solar system. The most massive known supermassive black holes exceed 10 billion solar masses, making them among the most massive single objects in the universe.

The formation of supermassive black holes remains one of astrophysics' greatest puzzles. They are far too massive to have formed from single stellar collapse, yet observations show they existed when the universe was less than a billion years old—apparently too early for them to have grown to such sizes through accretion alone. Current theories propose various formation mechanisms, including the direct collapse of primordial gas clouds, the growth of intermediate-mass seed black holes formed by the first generation of stars, or even more exotic processes involving dark matter concentrations in the early universe.

Between these extremes, intermediate-mass black holes with masses between 100 and 100,000 solar masses remain elusive and controversial. While theoretical models suggest they should exist, confirming their presence observationally has proven challenging. Some candidates have been proposed based on X-ray observations and gravitational wave detections, but the intermediate-mass black hole population remains poorly understood.

Feeding the Beast - Accretion and Energy Production

While black holes themselves emit no light, the matter falling into them can become the most efficient energy sources in the universe, converting up to 42% of infalling mass into radiant energy—compared to just 0.7% for nuclear fusion in stars. This extraordinary efficiency makes black holes the engines driving some of the cosmos' most spectacular phenomena, from quasars that outshine entire galaxies to the X-ray emissions that first revealed their existence.

The key to this energy production lies in the process of accretion, whereby material with angular momentum cannot fall directly into a black hole but instead forms a rotating disk as it spirals inward. These accretion disks represent one of the most extreme environments in the universe, where matter accelerates to significant fractions of the speed of light while being heated to temperatures of millions or even billions of degrees.

As material in the accretion disk loses angular momentum through viscous processes and magnetic interactions, it gradually spirals closer to the black hole. The gravitational energy released during this infall gets converted to heat through friction and magnetic reconnection events. The inner regions of the disk, where orbital velocities approach the speed of light and gravitational forces are strongest, become so hot that they radiate primarily in X-rays and gamma rays.

The structure of accretion disks depends critically on the rate at which material falls toward the black hole. At high accretion rates, the disk becomes geometrically thick and radiatively efficient, producing the brilliant emissions we observe in quasars and active galactic nuclei. These objects can outshine their entire host galaxies, with luminosities exceeding 10^{12} times that of our Sun. At lower accretion rates, the disk becomes geometrically thick but radiatively inefficient, with much of the gravitational energy being advected inward with the flow rather than radiated away.

This transition between radiatively efficient and inefficient accretion helps explain the dichotomy between active and normal galaxies. Most galaxies, including our own Milky Way, contain supermassive black holes that are currently in a quiescent state, accreting material at rates too low for efficient radiation. However, when these black holes encounter sufficient fuel—whether from stellar winds, tidal disruption events, or galaxy mergers—they can transition to active states, becoming visible as quasars or other forms of active galactic nuclei.

The accretion process also produces some of the universe's most dramatic phenomena in the form of relativistic jets. These narrow beams of particles accelerate to speeds approaching that of light, extending for thousands of light-years beyond the black hole itself. The mechanism for jet production remains an active area of research, but it likely involves the extraction of rotational energy from spinning black holes through magnetic field lines threaded through the accretion disk—a process known as the Blandford-Znajek mechanism.

Observing the Invisible - X-ray Windows and Relativistic Effects

Since black holes emit no light directly, astronomers must rely on their effects on surrounding matter and spacetime to study these objects. X-ray astronomy has proved particularly valuable for black hole studies because the hot, rapidly moving material in accretion disks radiates most intensely at X-ray wavelengths. These observations provide a window into the most extreme gravitational environments in the universe, where the effects of general relativity become clearly observable.

X-ray observations of black hole systems reveal a complex interplay between accretion physics and relativistic effects. The spectra of X-rays from accretion disks show characteristic features that allow astronomers to probe the spacetime geometry near the event horizon. Iron emission lines, produced when X-rays from the hot inner disk illuminate cooler material further out, become dramatically broadened and skewed by relativistic effects including Doppler shifting, gravitational redshift, and light bending.

These relativistic modifications of the observed radiation serve as powerful tests of general relativity in the strong-field regime. Material on the approaching side of the accretion disk appears brighter and bluer due to relativistic beaming and Doppler boosting, while material on the receding side appears dimmer and redder. The exact pattern of these asymmetries depends on the black hole's mass, spin, and the viewing angle of the system, allowing astronomers to measure these fundamental parameters.

Frame-dragging effects, predicted by general relativity for rotating black holes, have been observed in several systems. When a black hole spins, it literally drags spacetime around with it, causing the orbital plane of nearby material to precess. This effect has been dramatically observed in systems like V404 Cygni, where the jets from the black hole wobble rapidly as the inner accretion disk precesses due to frame-dragging. These observations provide direct confirmation of one of general relativity's most subtle predictions and demonstrate how black holes actively shape the spacetime around them.

The study of X-ray variability from black hole systems has revealed the chaotic nature of accretion flows and provided insights into the physics very close to the event horizon. Rapid flickering in X-ray brightness, occurring on timescales of milliseconds, indicates that the emission originates from regions just a few Schwarzschild radii from the black hole. The patterns of this variability carry information about the structure of the accretion flow and the strong gravitational fields that shape it.

Recent advances in X-ray polarimetry have opened new windows for studying black hole physics. The polarization of X-rays from accretion disks carries information about the magnetic field structure and the geometry of the emission region. These measurements help distinguish between different models for how jets are launched and how energy is extracted from rotating black holes.

Seeing the Shadow - The Event Horizon Telescope Revolution

The ultimate confirmation of black hole theory came through one of the most ambitious observational projects in astronomy: the Event Horizon Telescope (EHT), which succeeded in directly imaging the shadow of a black hole. This international collaboration created an Earth-sized virtual telescope by linking radio observatories across the globe, achieving the resolution necessary to peer into the immediate vicinity of a supermassive black hole's event horizon.

The EHT's first target was M87*, the supermassive black hole at the center of the giant elliptical galaxy Messier 87, located

55 million light-years from Earth. With a mass of 6.5 billion solar masses, M87* presented an ideal target because its enormous mass and relative proximity make it appear large enough in the sky to be resolved by even an Earth-sized telescope. The event horizon itself spans approximately 40 billion kilometers—roughly three times the size of Pluto's orbit—but appears as small as an orange on the Moon when viewed from Earth.

The image released in April 2019 revealed exactly what general relativity had predicted: a bright ring of emission surrounding a dark central region—the shadow of the black hole. This shadow represents not the event horizon itself, but rather the boundary between spacetime regions from which light can escape to reach distant observers and those from which it cannot. The observed shadow is approximately 2.5 times larger than the actual event horizon due to the bending of light by the black hole's gravity.

The asymmetric brightness of the ring provided crucial information about the black hole's properties. The southern portion appears brighter than the northern portion, indicating that material in the accretion disk is moving toward us on the south side and away from us on the north side. This pattern reveals that M87* is rotating clockwise as viewed from Earth and allows astronomers to determine the orientation of its spin axis relative to the massive jet that extends thousands of light-years from the galaxy's center.

Follow-up observations in 2018 confirmed the persistence of M87*'s shadow structure while revealing changes in the brightness distribution around the ring. These variations, occurring on timescales of about a year, match theoretical expectations for turbulent material orbiting near a black hole's event horizon. The consistency between observations separated by a year provided strong confirmation that the EHT was indeed imaging the immediate vicinity of the event horizon rather than some other phenomenon.

The EHT collaboration's second major achievement came with the imaging of Sagittarius A*, the supermassive black hole at the center of our own Milky Way galaxy. Despite being much closer to Earth at a distance of 26,000 light-years, Sgr A* presented unique observational challenges due to its smaller size and more rapidly varying emission. The resulting image, released in 2022, showed a remarkably similar ring structure to M87*, providing additional confirmation of general relativity's predictions and demonstrating that supermassive black holes share common properties regardless of their environment.

These breakthrough observations represent far more than mere confirmation of existing theories. They have opened an entirely new field of black hole studies, allowing astronomers to probe the strong-field regime of general relativity with unprecedented precision. Future EHT observations will measure how the shadow changes over time, potentially revealing the effects of orbiting material and testing general relativity's predictions about black hole spacetime with exquisite accuracy.

The success of the Event Horizon Telescope demonstrates the power of international scientific cooperation and technological innovation. By combining data from telescopes in locations ranging from Chile to the South Pole, the project achieved angular resolution that would be impossible with any single instrument. The massive computational challenges of processing and analyzing the data required innovative algorithms and supercomputing resources, pushing the boundaries of both astronomy and data science.

Looking ahead, plans for expanding the EHT network with additional telescopes and even space-based observations promise even sharper images and new capabilities for studying black hole physics. These future observations may finally resolve long-standing questions about how jets are launched, how information behaves near event horizons, and whether general relativity remains valid under the most extreme conditions the universe can provide.

Black holes have evolved from speculative concepts to directly observable objects, revealing themselves as laboratories for testing the fundamental laws of physics under conditions impossible to reproduce on Earth. They challenge our understanding of space, time, information, and quantum mechanics while powering some of the universe's most energetic phenomena. As we peer deeper into these cosmic abysses, we continue to discover that black holes are not just endpoints of stellar evolution or gravitational curiosities, but integral components of cosmic structure formation and galaxy evolution. They represent the ultimate intersection where Einstein's general relativity meets quantum mechanics, where information theory confronts thermodynamics, and where our most sophisticated theories of physics approach their limits. The journey from John Michell's dark stars to the Event Horizon Telescope's shadow images demonstrates both the power of theoretical physics to predict phenomena decades before observation and the ingenuity of observational astronomy to test the universe's most extreme predictions. In studying black holes, we study the fundamental nature of reality itself, exploring regions where physics meets its limits and discovers new possibilities.

Chapter 11: High-Energy Phenomena - The Universe's Most Violent Events

The universe's most spectacular displays of energy occur where matter encounters the most extreme physical conditions—regions of immense gravitational fields, crushing magnetic forces, and violent nuclear processes that accelerate particles to nearly the speed of light and generate radiation across the entire electromagnetic spectrum. These high-energy phenomena serve as cosmic laboratories where physics operates under conditions impossible to achieve on Earth, revealing fundamental processes that shape galactic evolution and cosmic structure. From gamma-ray bursts that release more energy in seconds than the Sun will produce over its entire lifetime, to supernova remnants that accelerate cosmic rays to energies millions of times greater than the most powerful terrestrial particle accelerators, these violent events demonstrate nature's extraordinary capacity for energy release and particle acceleration. Multi-wavelength observations have revolutionized our understanding of these phenomena, showing how the universe's most energetic processes are interconnected through complex feedback mechanisms that influence star formation, galactic chemical evolution, and the growth of cosmic structure itself.

The Nature of High-Energy Astrophysics

High-energy astrophysics encompasses the study of exceedingly dynamic and energetic phenomena occurring near the most extreme objects in the universe, including black holes, neutron stars, white dwarfs, and supernova remnants. These objects convert gravitational binding energy into the acceleration of extremely energetic particles called cosmic rays, which exhibit non-thermal energy spectra spanning from 10^{10} electron volts all the way up to ultra-high-energy cosmic rays with energies reaching 10^{20} electron volts. Such energies represent particles moving at velocities approaching the speed of light, carrying individual energies comparable to a baseball thrown at 100 kilometers per hour—concentrated in a single subatomic particle.

The extreme environments around compact objects create conditions where standard plasma physics breaks down and relativistic effects become dominant. Temperatures can reach billions of degrees, magnetic fields can exceed those of laboratory magnets by factors of trillions, and matter densities can approach those found in atomic nuclei. These conditions naturally produce radiation across the electromagnetic spectrum, from radio waves to gamma rays, with each wavelength revealing different aspects of the underlying physics. The multi-wavelength approach has become essential for understanding high-energy phenomena because no single wavelength can capture the full complexity of these systems.

X-ray observations have proved particularly valuable because the hot, rapidly moving material in these extreme environments radiates most intensely at X-ray wavelengths. These observations provide windows into the most extreme gravitational environments in the universe, where general relativistic effects become clearly observable and where matter behaves in ways that challenge our understanding of fundamental physics. The development of space-based X-ray telescopes has been crucial for this field, as Earth's atmosphere completely absorbs X-rays, making ground-based observations impossible.

Gamma-Ray Bursts - The Universe's Ultimate Explosions

Gamma-ray bursts represent the most energetic electromagnetic events known to science, capable of releasing energies between 10^{50} and 10^{54} ergs in gamma rays alone—energy outputs that can exceed what an entire galaxy produces in visible light. These mysterious phenomena were first discovered during the height of the Cold War, when American military satellites designed to monitor Soviet nuclear weapons testing instead detected powerful flashes of gamma rays coming from deep space. The Vela satellites, launched in the 1960s to detect gamma rays from nuclear explosions, began recording brief, intense bursts of high-energy radiation that clearly originated from beyond the solar system.

The discovery of gamma-ray bursts presented astronomers with one of their greatest mysteries. The bursts appeared randomly across the sky with no obvious pattern, lasting anywhere from milliseconds to several minutes before fading away. For decades, their origins remained completely enigmatic—were they from nearby neutron stars, distant galaxies, or something entirely different? The breakthrough came in 1997 when the BeppoSAX satellite managed to quickly localize a gamma-ray burst and alert ground-based telescopes, which detected a fading optical counterpart in a distant galaxy. This observation definitively established that gamma-ray bursts occur at cosmological distances, making them by far the most luminous events in the universe.

Further observations revealed that gamma-ray bursts fall into two distinct categories based on their duration and spectral properties. Long-duration bursts last more than two seconds and can continue for several minutes, while short-duration bursts typically last less than two seconds. This bimodal distribution suggested different physical mechanisms were responsible for each type, leading to intensive theoretical and observational efforts to understand their origins.

Collapsars and Long-Duration Bursts

Long-duration gamma-ray bursts are now understood to result from the catastrophic collapse of extremely massive stars, specifically Wolf-Rayet stars or other stars with initial masses exceeding 25-30 times that of the Sun. These stars have burned through their nuclear fuel so rapidly that their cores become unstable and collapse directly into black holes, bypassing the neutron star phase entirely. The key ingredient for producing a gamma-ray burst is rapid rotation—the collapsing stellar core must be spinning fast enough to form a black hole surrounded by a hot, rapidly accreting disk of stellar material.

The collapsar scenario explains how these stellar death throes can produce such extraordinary energy outputs. As the stellar core collapses and forms a black hole, the remaining stellar envelope continues to fall inward, forming a massive accretion disk around the newly formed black hole. The gravitational energy released as this material spirals into the black hole powers the generation of powerful, highly collimated jets of relativistic plasma along the rotational axis of the system. These jets must punch through thousands of kilometers of stellar material before breaking free into space, a process that takes several seconds and produces the characteristic long duration of these bursts.

Recent simulations have shown that the process is remarkably violent and complex. The jets interact with the collapsing stellar envelope, creating shock waves and turbulent mixing that affect both the energy output and the spectral characteristics of the resulting gamma-ray burst. The jets themselves achieve Lorentz factors of 10 or higher, meaning they travel at velocities exceeding 99.5% of the speed of light. When these ultra-relativistic jets break free from the stellar envelope, they produce the observed gamma-ray emission through a combination of internal shocks within the jet and external shocks as the jet interacts with the surrounding interstellar medium.

The collapsar model also explains why long gamma-ray bursts are associated with star-forming regions and why they show evidence of heavy element enrichment. The massive progenitor stars that produce these events have short lifetimes and therefore explode in or near the star-forming regions where they were born. Furthermore, these explosions likely trigger additional star formation in nearby molecular clouds through the shock waves they generate, creating a feedback mechanism that influences galactic evolution.

Short Bursts and Neutron Star Mergers

Short-duration gamma-ray bursts have been linked to an entirely different class of events: the merger of binary neutron star systems or the merger of a neutron star with a black hole. This connection was dramatically confirmed on August 17, 2017, when the LIGO and Virgo gravitational wave detectors recorded the collision of two neutron stars in the galaxy NGC 4993, located about 140 million light-years from Earth. Just 1.7 seconds after the gravitational wave signal ended, the Fermi Gamma-ray Space Telescope detected a short gamma-ray burst from the same region of sky.

This simultaneous detection of gravitational waves and gamma rays provided definitive proof that at least some short gamma-ray bursts originate from neutron star mergers. The gravitational wave signal GW170817 showed the characteristic frequency evolution expected from two objects with masses consistent with neutron stars, spiraling inward over the final 100 seconds before merger. The associated gamma-ray burst, designated GRB 170817A, was much weaker than typical short bursts, but its temporal and spectral characteristics were consistent with theoretical models of jet emission from neutron star mergers.

The multi-messenger observations of this event revealed extraordinary details about the merger process. Ground-based telescopes detected optical and infrared emission beginning about 11 hours after the gravitational wave detection, marking the first observation of a kilonova—the electromagnetic counterpart to a neutron star merger. This emission came from the radioactive decay of heavy elements synthesized in the neutron-rich material ejected during the merger, providing direct evidence that neutron star collisions are major sites of r-process nucleosynthesis.

The kilonova associated with GW170817 evolved exactly as theoretical models predicted. The early emission was blue and peaked in the ultraviolet and optical wavelengths, consistent with the decay of light r-process elements like strontium and yttrium. Over the following days, the emission became increasingly red as heavier lanthanide elements began to dominate the radioactive energy release. Spectroscopic observations confirmed the presence of strontium and likely other heavy elements, proving that these mergers produce significant quantities of elements that cannot be formed in ordinary stellar nucleosynthesis.

X-Ray Astronomy and Accretion Physics

X-ray astronomy has opened unique windows into high-energy processes throughout the universe, revealing phenomena invisible at other wavelengths. Unlike optical light, X-rays are produced primarily by extremely hot gas, typically with temperatures exceeding millions of degrees, or by highly energetic non-thermal processes involving relativistic particles. The development of focusing X-ray telescopes and high-resolution spectrometers has revolutionized our ability to study these extreme environments with unprecedented detail.

Accretion-powered X-ray sources represent one of the most important classes of high-energy phenomena. When matter falls toward a compact object such as a neutron star or black hole, the gravitational potential energy released can be converted to X-ray radiation with remarkable efficiency. For material accreting onto a neutron star, conversion efficiencies can reach 10-20%, while accretion onto black holes can achieve efficiencies approaching 42% for maximally spinning black holes. These efficiencies far exceed those of nuclear fusion, making accretion the most efficient energy source known in the universe.

The structure and dynamics of accretion flows depend critically on the properties of the compact object and the rate at which material is being accreted. At high accretion rates, the flow becomes geometrically thick and radiatively efficient, producing intense X-ray emission with characteristic spectral features that reveal the physical conditions near the compact object. X-ray spectroscopy has become a powerful probe of strong-field gravity, as the radiation from these systems is modified by relativistic effects including Doppler shifting, gravitational redshift, and light bending.

Recent X-ray observations have confirmed many predictions of general relativity while revealing unexpected complexities in accretion physics. Iron emission lines from accretion disks show characteristic broadening and asymmetries caused by the extreme gravitational fields near black holes and neutron stars. These features allow astronomers to measure fundamental properties of compact objects, including their masses, spins, and the structure of spacetime in their immediate vicinity. Some systems show evidence for magnetic field reconnection and other plasma instabilities that affect the observed X-ray variability.

Frame-Dragging and Relativistic Effects

One of the most remarkable confirmations of general relativity in high-energy astrophysics comes from observations of frame-dragging effects around rotating black holes. When a black hole spins, it literally drags the surrounding spacetime around with it, causing nearby matter to be forced into co-rotation. This phenomenon, also known as the Lense-Thirring effect, has been directly observed through the behavior of X-ray emission from accretion disks and the precession of relativistic jets.

In several black hole X-ray binary systems, astronomers have observed rapid wobbling of jets that can only be explained by frame-dragging. The inner regions of the accretion disk are forced to precess around the black hole's spin axis, and this precession is transmitted to the jets that are launched from the disk. The timescales of this precession provide direct measurements of the black hole's spin and confirm the predictions of general relativity in the strong-field regime.

Frame-dragging effects are most pronounced near rapidly spinning black holes, where they can be observed through several different mechanisms. High-resolution X-ray spectroscopy can detect the orbital motions of material in the inner accretion disk, revealing the twisted geometry of spacetime around spinning black holes. Polarization measurements of X-rays can also reveal frame-dragging through the way magnetic fields are dragged around with the spacetime flow. These observations represent some of the most direct tests of Einstein's theory under the most extreme conditions nature provides.

Supernova Remnants as Cosmic Accelerators

Supernova remnants serve as the galaxy's primary factories for high-energy cosmic rays, accelerating particles to energies that exceed those achievable in the most powerful terrestrial particle accelerators. When a massive star explodes as a supernova, it creates a shock wave that expands into the surrounding interstellar medium at velocities of thousands of kilometers per second. These shocks can persist for tens of thousands of years, providing the conditions necessary for efficient particle acceleration through a process known as diffusive shock acceleration.

The mechanism for cosmic ray acceleration in supernova remnants was first proposed by Enrico Fermi in 1949, but direct observational evidence remained elusive for decades. The breakthrough came with gamma-ray observations from NASA's Fermi Gamma-ray Space Telescope, which detected the characteristic signature of high-energy protons interacting with interstellar gas in two particular supernova remnants: IC 443 (the Jellyfish Nebula) and W44. These observations provided the first definitive "smoking gun" evidence that supernova remnants do indeed accelerate cosmic ray protons to high energies.

The acceleration process occurs when charged particles get trapped in the magnetic fields associated with the supernova shock wave. As particles cross back and forth through the shock front, they gain energy with each crossing—typically about 1% per crossing. After many such crossings, which can occur over thousands of years, particles can achieve energies millions of times higher than their initial thermal energies. The most energetic particles eventually escape the remnant and propagate through the galaxy as cosmic rays.

Recent theoretical work has shown that supernova remnants can potentially accelerate particles to "knee" energies around 10^{15} electron volts, but only under specific conditions. The key requirements include high shock velocities exceeding 10,000 kilometers per second and significant magnetic field amplification driven by the cosmic rays themselves. Young supernova remnants expanding into dense stellar winds may represent the best candidates for achieving these extreme energies.

Magnetic Field Amplification and Particle Confinement

The efficiency of cosmic ray acceleration in supernova remnants depends critically on the strength of magnetic fields in the shock region. In the standard interstellar medium, magnetic field strengths are only a few microGauss, which would be insufficient to confine and accelerate particles to the observed cosmic ray energies. However, the cosmic rays themselves can drive instabilities that amplify the magnetic field by factors of 100 or more.

This magnetic field amplification occurs through a process called the non-resonant hybrid instability, which is triggered by cosmic rays escaping from the shock region. As high-energy particles stream away from the shock, they create electric currents in the plasma that destabilize the magnetic field configuration. The resulting magnetic turbulence can grow rapidly, creating the strong fields necessary to confine particles near the shock and enable further acceleration.

Observations of supernova remnants support this theoretical picture. X-ray observations reveal thin filaments of emission that indicate magnetic fields have been compressed and amplified by factors of 100 or more compared to the background interstellar field. The time evolution of these features matches theoretical predictions for magnetic field amplification driven by cosmic ray acceleration. However, the process appears to be limited by the age of the supernova remnant, as the magnetic amplification becomes less efficient as the shock velocity decreases over time.

Multi-Wavelength Signatures of Acceleration

The acceleration of cosmic rays in supernova remnants produces observable signatures across the electromagnetic spectrum. At radio wavelengths, synchrotron radiation from accelerated electrons creates the characteristic shell-like morphology observed in many supernova remnants. The radio spectral properties provide information about the energy spectrum of the accelerated electrons and the strength of magnetic fields in the emission region.

Gamma-ray emission provides the most direct probe of cosmic ray proton acceleration. When high-energy protons interact with ambient gas, they produce neutral pions that decay into gamma rays with energies exceeding 100 MeV. The detection of this "pion bump" in the gamma-ray spectrum provides unambiguous evidence for cosmic ray proton acceleration, as electrons cannot produce gamma rays at these energies through any plausible mechanism.

X-ray observations reveal the thermal properties of the shocked gas and can also detect synchrotron emission from the highest-energy electrons. The combination of thermal and non-thermal X-ray emission provides constraints on the efficiency of particle acceleration and the fraction of the shock energy that goes into accelerating cosmic rays rather than heating the gas. These multi-wavelength observations have shown that the most efficient cosmic ray accelerators are typically young supernova remnants with high shock velocities expanding into dense circumstellar material.

Ultra-High-Energy Cosmic Rays and Extragalactic Sources

While supernova remnants can explain the galactic cosmic ray population up to energies around 10^{15} electron volts, the highest-energy cosmic rays—those with energies exceeding 10^{18} electron volts—require more exotic acceleration sites. These ultra-high-energy cosmic rays likely originate from extragalactic sources, possibly including active galactic nuclei, gamma-ray bursts, or other extreme phenomena associated with supermassive black holes.

The propagation of ultra-high-energy cosmic rays through intergalactic space is complicated by their interactions with the cosmic microwave background radiation. Protons with energies above about $6\text{x}10^{19}$ electron volts interact with microwave photons through photopion production, limiting the distance from which such particles can reach Earth. This "GZK cutoff" means that the highest-energy cosmic rays must originate from relatively nearby sources within about 50 Megaparsecs.

Recent theoretical work suggests that cosmic rays may be self-confined near their extragalactic sources through plasma instabilities similar to those operating in supernova remnants. The large electric currents associated with cosmic ray outflows from powerful sources can drive magnetic field amplification that traps particles in the source environment for extended periods. This self-confinement may help explain the observed spectrum and composition of ultra-high-energy cosmic rays while providing additional time for particles to reach extreme energies.

The study of high-energy phenomena continues to reveal new aspects of cosmic violence and energy release. From the catastrophic deaths of the most massive stars in gamma-ray bursts to the steady acceleration of particles in supernova shocks, these processes demonstrate the universe's extraordinary capacity for creating extreme physical conditions. As new generations of instruments come online—including next-generation gravitational wave detectors, advanced gamma-ray telescopes, and more sensitive cosmic ray observatories—we can expect continued breakthroughs in understanding these spectacular displays of cosmic energy. The multi-messenger approach, combining gravitational waves, electromagnetic radiation, and particle detections, promises to reveal even deeper connections between these phenomena and their roles in shaping the cosmic environment. These violent events are not merely spectacular displays of nature's power, but fundamental processes that drive galactic evolution, create the heavy elements necessary for planet formation, and maintain the dynamic equilibrium that makes the universe the vibrant, evolving system we observe today.

Chapter 12: The Cosmic Symphony - Connecting the Universe's Greatest Stories

The journey through astrophysics reveals a universe of breathtaking complexity and elegant simplicity, where the same fundamental physical laws that govern a falling apple on Earth also determine the fate of galaxies billions of light-years away. As we reach the conclusion of our exploration, we can step back and appreciate how the diverse phenomena we have encountered—from the quantum fluctuations that seeded cosmic structure to the gravitational waves rippling through spacetime—form a interconnected narrative that tells the grand story of cosmic evolution. This is not merely a collection of isolated phenomena, but rather a cosmic symphony where each component plays its part in a larger composition that has been unfolding for 13.8 billion years. The universe emerges as a dynamic, evolving entity where matter and energy dance to the rhythms set by fundamental forces, creating the rich tapestry of structures and processes we observe today. Through our journey, we have discovered that astrophysics is ultimately about connections—connections between the very small and the unimaginably large, between the earliest moments of cosmic time and the present day, and between phenomena that initially seem unrelated but reveal deep underlying unity when viewed through the lens of physical law.

The Universal Clock - Cosmic Time as the Great Unifier

The concept of cosmic time emerges as one of the most profound organizing principles in our understanding of the universe, providing a universal clock that synchronizes events across vast distances and connects phenomena separated by billions of years. Unlike our everyday experience of time as a local phenomenon, cosmic time is intimately connected to the expansion of the universe itself, with the cosmic microwave background serving as a snapshot frozen in time from when the universe became transparent 380,000 years after the Big Bang. This ancient radiation, now cooled to a mere 2.7 Kelvin above absolute zero, carries within its minute temperature fluctuations the seeds of all cosmic structure that would later emerge.

The beauty of cosmic time lies in how it allows us to trace the universe's evolution through distinct epochs, each building upon the foundations laid by previous eras. During the first few minutes after the Big Bang, when temperatures exceeded billions of degrees, primordial nucleosynthesis forged the light elements that would later fuel the first stars. This process established the cosmic abundance ratio of roughly three parts hydrogen to one part helium that persists to this day, demonstrating how the universe's earliest moments continue to influence its present state.

The era from nucleosynthesis to recombination—when electrons first combined with atomic nuclei to form stable atoms—represents a crucial transition that fundamentally changed the universe's character. Before recombination, the cosmos was opaque to light, with photons constantly scattering off free electrons in a cosmic fog. The formation of neutral atoms at a cosmic age of 380,000 years suddenly made the universe transparent, allowing light to travel freely across cosmic distances for the first time. This transition not only produced the cosmic microwave background we observe today but also enabled the gravitational collapse that would create the first stars and galaxies.

The James Webb Space Telescope has revolutionized our ability to peer back into cosmic time, observing galaxies as they appeared when the universe was less than 5% of its current age. These observations reveal that galaxy formation occurred more rapidly and efficiently than previously thought, with massive galaxies already in place within the first billion years after the Big Bang. The discovery of "red monster" galaxies—objects nearly as massive as the Milky Way existing when the universe was extremely young—challenges our models of how quickly cosmic structure can assemble and suggests that the early universe was even more dynamic than we previously imagined.

Stellar Engines - The Cosmic Cycle of Creation and Destruction

Throughout our journey, stellar evolution has emerged as the central process driving cosmic chemical evolution and powering much of the universe's observable phenomena. Stars represent far more than distant points of light—they are cosmic factories that transform the universe's primordial hydrogen and helium into the periodic table of elements that make planets, life, and ultimately human beings possible. This stellar alchemy operates through nuclear fusion in stellar cores, where temperatures and pressures reach extreme values that force atomic nuclei to overcome their mutual electrical repulsion and combine into heavier elements.

The mass of a star at birth determines its entire life story, from its main sequence lifetime to its ultimate fate as a white dwarf, neutron star, or black hole. Low-mass stars like our Sun burn hydrogen slowly and steadily for billions of years before gently expelling their outer layers as planetary nebulae, leaving behind white dwarf cores that will cool slowly over cosmic ages. These white dwarfs, supported by electron degeneracy pressure, represent one endpoint of stellar evolution and serve as cosmic timepieces whose cooling rates help astronomers measure the ages of galactic populations.

More massive stars lead dramatically different lives, burning through their nuclear fuel rapidly and ending in spectacular supernova explosions that can briefly outshine entire galaxies. These stellar deaths are not merely endpoints but crucial events that enrich the interstellar medium with heavy elements while triggering shock waves that can compress nearby gas clouds and stimulate new star formation. The supernova process thus represents both destruction and creation, ending one stellar generation while preparing the materials and conditions for the next.

The most massive stars bypass the white dwarf and even neutron star phases, collapsing directly into black holes when their nuclear fuel is exhausted. These stellar-mass black holes, typically containing between 3 and 50 solar masses, demonstrate how gravity ultimately triumphs over all other forces when sufficient mass is concentrated in a small region. Yet even these cosmic monsters play constructive roles in galactic evolution, powering energetic phenomena through accretion processes and participating in the merger events that produce gravitational waves detectable across the universe.

Gravity's Cosmic Dominance - From Quantum Fluctuations to Black Holes

The role of gravity throughout cosmic history demonstrates how the weakest of the four fundamental forces ultimately shapes the universe's large-scale structure and evolution. Einstein's general theory of relativity reveals gravity not as a force in the conventional sense, but as the curvature of spacetime itself—a geometric property that determines how matter and energy move through the cosmos. This geometric understanding of gravity has proven remarkably powerful, with Einstein's field equations successfully describing phenomena from GPS satellite corrections to the most extreme environments near black hole event horizons.

The story of cosmic structure formation begins with tiny quantum fluctuations in the early universe that were stretched to macroscopic scales during the epoch of cosmic inflation. These fluctuations provided the seeds for all subsequent structure formation, with regions of slightly higher density gradually attracting more matter through gravitational attraction. Dark matter played a crucial role in this process because it responds only to gravity, allowing structures to begin forming earlier than would have been possible with ordinary matter alone.

The hierarchical nature of structure formation—where smaller structures merge to form larger ones—demonstrates how gravity operates as a cosmic organizer across all scales. Dark matter halos formed first, providing the gravitational scaffolding that later attracted ordinary matter to form the first stars and galaxies. This process continues today as galaxies merge to form larger structures, ultimately creating the cosmic web of galaxy clusters and superclusters separated by vast voids that characterizes the universe on its largest scales.

Black holes represent gravity's ultimate expression, where spacetime curvature becomes so extreme that not even light can escape once it crosses the event horizon. The recent direct imaging of black hole shadows by the Event Horizon Telescope has provided dramatic confirmation of general relativity's predictions about these exotic objects, showing how they warp the very fabric of space and time around them. These observations reveal black holes not as cosmic vacuum cleaners indiscriminately devouring everything nearby, but as dynamic objects that interact with their environments through accretion processes and jet formation.

The Dark Universe - Unseen Components Driving Cosmic Evolution

Perhaps the most humbling discovery in modern astrophysics has been the recognition that ordinary matter—the atoms that compose stars, planets, and human beings—represents only about 5% of the universe's total mass-energy content. The remaining 95% consists of dark matter and dark energy, mysterious components that reveal themselves only through their gravitational effects and their influence on cosmic expansion. This realization has fundamentally changed our understanding of the universe's composition while highlighting how much we still don't know about the cosmos we inhabit.

Dark matter's role in cosmic structure formation cannot be overstated. Without this invisible component, galaxies would not have had time to form by gravitational collapse within the age of the universe, and the cosmic web of large-scale structure we observe today would not exist. Computer simulations show that dark matter forms a complex network of filaments and halos that provide the gravitational framework for ordinary matter to collect and eventually form stars and galaxies. The precise nature of dark matter particles remains one of astrophysics' greatest unsolved mysteries, with detection experiments searching for everything from weakly interacting massive particles to axions to primordial black holes.

Dark energy presents an even more profound puzzle. Its discovery through observations of distant supernovae revealed that the universe's expansion is not just continuing but actually accelerating, contrary to expectations that gravity should gradually slow the expansion over time. This acceleration implies the existence of a form of energy that produces repulsive gravitational effects, driving space itself to expand at an ever-increasing rate. Whether dark energy represents Einstein's cosmological constant, a dynamic scalar field, or something even more exotic remains an open question that will likely require new theoretical breakthroughs to resolve.

The interplay between dark matter and dark energy determines the universe's ultimate fate. Current observations suggest we live in a universe dominated by dark energy, which will continue driving accelerated expansion into the far future. This scenario leads to a "heat death" where galaxies move ever farther apart, stars exhaust their fuel, and the universe gradually cools toward absolute zero over timescales that dwarf the current age of the cosmos. Understanding these dark components thus addresses fundamental questions about not only how the universe evolved but also where it is headed.

Multi-Messenger Revolution - New Windows on Cosmic Phenomena

The development of multi-messenger astronomy represents one of the most significant advances in our ability to study cosmic phenomena, combining gravitational waves, electromagnetic radiation, neutrinos, and cosmic rays to gain comprehensive understanding of the universe's most energetic events. This approach has already produced remarkable discoveries and promises to revolutionize astrophysics in the coming decades as new detectors come online and detection techniques improve.

The historic detection of gravitational waves by LIGO in 2015 opened an entirely new window on the universe, allowing us to directly observe the ripples in spacetime predicted by Einstein's general relativity. The first detection, from the merger of two black holes each about 30 times more massive than the Sun, confirmed not only the existence of gravitational waves but also revealed that binary stellar-mass black holes are common enough to produce detectable merger events regularly. The energy released in this single event—equivalent to converting three solar masses directly into gravitational wave energy—exceeded the combined light output of all stars in the observable universe for a few milliseconds.

Even more revolutionary was the joint detection of gravitational waves and electromagnetic radiation from the neutron star merger GW170817 in August 2017. This event provided direct confirmation that short gamma-ray bursts originate from neutron star collisions while revealing the process of r-process nucleosynthesis that creates many of the heaviest elements in the periodic table. The kilonova emission associated with this merger showed the characteristic spectral signatures of newly synthesized elements, demonstrating that neutron star mergers are cosmic foundries where gold, platinum, and other precious metals are created.

Future gravitational wave detectors, including the proposed Einstein Telescope and Cosmic Explorer, will provide orders-of-magnitude improvements in sensitivity and frequency coverage. These next-generation instruments will be capable of detecting neutron star mergers throughout the observable universe and may reveal entirely new classes of gravitational wave sources, from cosmic strings to phase transitions in the early universe. The combination of improved gravitational wave detection with electromagnetic follow-up observations using facilities like the James Webb Space Telescope will create unprecedented opportunities to study extreme physics under conditions impossible to reproduce in terrestrial laboratories.

Fundamental Connections - Recurring Themes Across Cosmic Scales

Throughout our exploration of astrophysical phenomena, certain fundamental themes have emerged repeatedly, demonstrating deep connections between apparently disparate processes operating across vastly different scales and environments. These recurring patterns reveal the underlying unity of physical law and show how the same basic principles manifest in diverse cosmic contexts.

Magnetic fields represent one such unifying theme, playing crucial roles in phenomena ranging from stellar formation to galaxy-scale jets launched by supermassive black holes. In star-forming regions, magnetic fields help regulate the collapse of molecular clouds and influence the initial mass function of newly formed stars. Around compact objects, magnetic fields channel accreting material into powerful jets that can extend for thousands of light-years, while in supernova remnants, magnetic field amplification enables the acceleration of cosmic rays to extraordinary energies. The tangled magnetic field lines that power cosmic particle accelerators demonstrate how magnetic energy can be converted into kinetic energy with remarkable efficiency.

Angular momentum conservation emerges as another fundamental principle that shapes cosmic evolution across all scales. From the formation of planetary systems through the collapse of rotating molecular clouds to the spin-up of neutron stars during core collapse supernovae, the tendency to conserve angular momentum determines system structure and evolution. Recent observations show that galaxies with high specific angular momentum tend to form stable disk structures with spiral patterns, while those with low angular momentum become turbulent and clumpy systems. This connection between angular momentum and galaxy morphology demonstrates how fundamental physical principles operating during galaxy formation continue to influence galactic properties billions of years later.

The tension between gravitational attraction and various forms of pressure—whether radiation pressure in stellar cores, degeneracy pressure in white dwarfs and neutron stars, or dark energy's repulsive effects on cosmic scales—represents a third recurring theme that determines the structure and evolution of cosmic systems. This balance manifests differently depending on the system's mass, composition, and environment, but the underlying physics remains the same. Understanding how these competing effects operate across different regimes has allowed astronomers to predict everything from stellar lifetimes to the ultimate fate of the universe.

Future Frontiers - The Next Chapter in Cosmic Discovery

As we look toward the future of astrophysical research, several key questions and technological developments promise to drive major advances in our understanding of the cosmos. The resolution of these questions will likely require both observational breakthroughs and theoretical innovations that may fundamentally alter our picture of the universe.

The nature of dark matter and dark energy remains the most pressing challenge in cosmology. Ground-based experiments continue searching for dark matter particles through direct detection, while space-based missions map the cosmic web's structure to constrain dark energy models. The upcoming Euclid mission and the Nancy Grace Roman Space Telescope will provide unprecedented precision measurements of cosmic expansion and structure formation, potentially revealing whether dark energy evolves over time or represents a true cosmological constant. Resolution of these mysteries may require new physics beyond the Standard Model of particle physics, possibly involving extra dimensions, modified gravity theories, or exotic forms of matter and energy.

Quantum gravity effects near black holes represent another frontier where observational capabilities are finally approaching theoretical requirements. The Event Horizon Telescope's direct imaging of black hole shadows has provided the first direct tests of spacetime structure in the strong-field regime, while gravitational wave detections probe the dynamics of black hole mergers with exquisite precision. Future developments in quantum gravity theory, combined with next-generation gravitational wave detectors and higher-resolution black hole imaging, may finally reveal how quantum mechanics and general relativity merge at the most extreme energy scales.

The formation of the first stars and galaxies continues to push observational capabilities to their limits. The James Webb Space Telescope has already revolutionized our understanding of early galaxy formation by discovering massive galaxies existing much earlier than theoretical models predicted. Future extremely large telescopes, including the Thirty Meter Telescope and the European Extremely Large Telescope, will provide even greater sensitivity and resolution for studying the universe's first billion years. These observations may reveal whether the first stars were fundamentally different from those forming today and how supermassive black holes grew so rapidly in the early universe.

The Cosmic Perspective - Humanity's Place in the Universal Story

The study of astrophysics provides more than just knowledge about distant objects and processes—it fundamentally transforms our understanding of humanity's place within the cosmic story. This cosmic perspective reveals both our apparent insignificance in terms of spatial and temporal scales and our remarkable significance as conscious beings capable of comprehending the universe's grand design. The recognition that we are composed of elements forged in stellar cores gives literal meaning to the phrase "we are made of star stuff," connecting human existence to cosmic processes operating across billions of years.

The anthropic principle highlights the apparent fine-tuning of physical constants that makes complex structure formation possible. Small changes in fundamental parameters like the strength of nuclear forces or the mass of elementary particles would prevent the formation of atoms, stars, or planets as we know them. This observation has led to various interpretations, from multiverse theories where we exist in one of countless universes with different physical laws, to more philosophical reflections on the relationship between observer and observed in scientific inquiry.

The universality of physical law demonstrated throughout astrophysics provides profound insights into the nature of scientific knowledge itself. The fact that equations developed to understand terrestrial phenomena accurately describe the behavior of objects billions of light-years away suggests deep underlying principles governing reality. This universality implies that the scientific method, developed through human experience on Earth, can legitimately be applied to understand phenomena far beyond direct observation or experimentation.

Modern astrophysics also reveals the universe as fundamentally dynamic and evolving rather than static and eternal. From the cosmic microwave background's tiny temperature fluctuations to the accelerating expansion driven by dark energy, we observe a cosmos in constant change and development. This recognition has philosophical implications for how we understand time, causation, and our own place within an evolving cosmic story that stretches from the Big Bang to an unimaginably distant future.

The remarkable success of astrophysics in explaining cosmic phenomena demonstrates humanity's extraordinary intellectual capabilities while simultaneously revealing the vast extent of what remains unknown. Each discovery answers existing questions while raising new ones, maintaining the sense of mystery and wonder that originally drew humans to study the night sky. This ongoing dialogue between theory and observation continues to drive scientific progress while inspiring new generations to explore the cosmos.

Synthesis - The Unity Behind Cosmic Diversity

As we conclude our journey through astrophysics, we can appreciate how the universe's apparent complexity emerges from the interplay of relatively simple fundamental principles operating across vast ranges of scale and time. The same physical laws that govern atomic behavior in terrestrial laboratories also determine stellar structure, galactic dynamics, and cosmological evolution. This unity underlying cosmic diversity represents one of science's most profound achievements—the recognition that natural phenomena, however exotic they may appear, operate according to comprehensible physical principles.

The story that emerges from modern astrophysics is one of cosmic evolution driven by the fundamental forces of nature, where gravity's attractive influence competes with various forms of pressure and repulsion to create the rich structures we observe. From the initial quantum fluctuations that seeded cosmic structure to the ongoing acceleration of cosmic expansion, the universe reveals itself as a dynamic system governed by physical law rather than arbitrary chance or supernatural intervention.

Yet for all our advances in understanding, astrophysics continues to confront us with profound mysteries that challenge our theoretical frameworks and observational capabilities. The dominance of dark matter and dark energy in cosmic evolution, the nature of quantum gravity effects near black holes, and the origin of cosmic structure all represent frontiers where current knowledge reaches its limits. These mysteries ensure that astrophysics will remain an active and exciting field of research for generations to come.

The development of multi-messenger astronomy and next-generation observational facilities promises continued breakthroughs that will deepen our understanding while likely revealing entirely new phenomena. The universe has consistently surprised astronomers with discoveries that were unanticipated by existing theories, from the cosmic microwave background to dark energy to gravitational waves. This pattern suggests that future observations will continue expanding our cosmic perspective in ways we cannot currently imagine.

Astrophysics ultimately represents humanity's most ambitious intellectual endeavor—the attempt to understand the universe as a whole through the application of physical principles discovered through terrestrial experience. The remarkable success of this enterprise demonstrates both the universality of natural law and the extraordinary power of human reason to comprehend phenomena far beyond direct sensory experience. From ancient observations of planetary motions to modern detections of gravitational waves from colliding black holes, the progressive refinement of our cosmic understanding shows how scientific knowledge builds cumulatively over time while remaining always open to revision and improvement.

The cosmic symphony we have explored reveals a universe of stunning beauty and profound mystery, where matter and energy dance to rhythms set by fundamental forces across scales from the subatomic to the cosmic. As conscious beings capable of contemplating our cosmic context, we occupy a unique position in this grand narrative—simultaneously products of stellar evolution and observers capable of understanding the processes that created us. This dual role as both participants in and students of cosmic evolution provides the ultimate motivation for continued exploration of the universe's greatest stories, ensuring that the cosmic symphony will continue to inspire wonder and drive discovery for as long as human consciousness persists.